STEM Education for the 21st Century

Bryan Edward Penprase

STEM Education for the 21st Century

Springer

Bryan Edward Penprase
Dean of Faculty
Soka University of America
Aliso Viejo, CA, USA

ISBN 978-3-030-41632-4 ISBN 978-3-030-41633-1 (eBook)
https://doi.org/10.1007/978-3-030-41633-1

This Springer imprint is published by the registered company Springer Nature Switzerland AG
The registered company address is: Gewerbestrasse 11, 6330 Cham, Switzerland

Preface: STEM in a New Age

This book arises from an odyssey through the world of higher education which took me across the world and across USA to visit leading universities, technology centers, and to meet with the top STEM educators who together are shaping a "new age" in STEM education. After teaching physics and astronomy for 20 years at Pomona College, conducting astrophysics research at Caltech, and leading several initiatives for online learning, curriculum development, and faculty training, including helping to found a new liberal arts college in Singapore, I wanted to document some of the vistas I saw while interviewing and working with visionary educators who are reshaping STEM education today. The journey included a year as a fellow with the American Council on Education, based at Yale University, which enabled me to visit Presidents, Provosts, Deans, Directors, and Professors across the world who are developing exciting new curricula, pedagogies, and technologies to help improve STEM education. In the ACE fellowship, I visited over 50 institutions of higher education around the world, and met with the founders and developers of new centers for online learning and entirely new campuses which were being developed to offer more efficient and effective forms of STEM education.

The combination of insights convinced me that a revolution is happening in STEM education which arises from a convergence of educational research, new technologies, and new ways of structuring both the physical space and classroom activities in higher education. Active learning, scientific teaching, and other techniques have enabled faculty to systematically improve learning for students across diverse populations, and the book describes some of the ways that leaders in STEM education have pioneered and refined these techniques in the past decade. The book is an ideal starting point for any concerned citizen or faculty member interested in getting an in-depth look at the most exciting new developments in STEM teaching and learning at our leading universities, and features first-hand accounts of how the development of entirely new types of curriculum and pedagogy, with interviews from top scientists across the world. The book also provides an inside look at the leading centers of online learning and engineering, and what the evolution of these transformative technologies will mean for our future. The book would be useful for any interested citizen who wants to see higher education is adapting to prepare students for a world transformed by science and technology, and also will equip any faculty member interested in contributing to the exciting "new age" within STEM education with a wealth of resources and specific strategies for curriculum development and teaching.

In separate chapters, the book tackles the most important issues facing STEM education today. Beginning with a chronicle of the history of STEM education in the US, the book makes the case for STEM education to include more diverse populations from across all socioeconomic, racial, and gender categories as a necessary phase in the long history of advancing social mobility through access to higher education in the US. A first-hand account of the development of revolutionary STEM curriculum and pedagogy includes detailed accounts of new programs that bring research into the classroom, interviews with top educators in biology, physics, and engineering, and an in-depth study of Nobel laureate Carl Wieman's science education initiative. The book offers specific and easy-to-use steps for implementing active and peer learning in the classroom, and bases these techniques within the context of educational research and theories of learning with interviews and quotes from leading innovators in STEM education. The book explores the latest conceptual frameworks for course design and learning, and explains Scientific Teaching, Backward Course Design, Learning Taxonomies, and Threshold Concepts in a systematic and easy-to-use way. As STEM enters this new age, these new theories and taxonomies of learning are guiding educators in developing their more effective courses and the book explores how our improved understanding of how people learn can enable STEM courses to be more effective. Within each chapter, the book provides dozens of useful links and resources, to help educators learn more and adapt some of the specific ideas to their teaching.

The book also takes the reader inside the world's most innovative new engineering programs, including Yale Engineering, with its Center for Engineering Innovation and Design, Olin College of Engineering, a new "startup" engineering school, and Cal Poly San Luis Obispo, with its approach of "learning by doing." The rise of new online courses, MOOCs, SPOCs, and blended learning, have upended many of our paradigms of higher education. The book takes the reader into the leading centers for online education, with site visits and interviews with the founders of Udacity, Coursera, and edX, who describe their vision of online education, and how these new technologies will evolve to transform STEM education. The global dimension of STEM education is also included, with research on global interdisciplinary science programs that includes interviews with 13 leaders of interdisciplinary STEM in six countries, with detailed insights into each of these innovative programs, and research on interdisciplinary theory, curriculum integration, and best practices for interdisciplinary STEM education. The book provides an expansive overview for anyone interested in the rise of online learning and global science curriculum, as well as an entry point for any educator, interested in adopting these new research-validated approaches to teaching science, engineering, and mathematics.

As advances in Artificial Intelligence, Biotechnology, and Materials Science transform our world through what some of called a "Fourth Industrial Revolution," citizens and educators need to be prepared for the ways that this will radically challenge STEM curriculum and high education itself. The book provides a rallying call for anyone interested in helping higher education respond to these "exponential technologies" and help STEM education incorporate the advances in learning science and technology to help us develop a more sustainable and peaceful future. From this new age comes the opportunity to reshape higher education institutions to enable them to implement more effective STEM education for all students, and to improve access to this education by a broader cross-section of society—to enable more of our young people to fully develop their talents and skills to help us reshape the future of our world.

Aliso Viejo, California, USA Bryan Edward Penprase

Acknowledgments

I would like to thank my mentors and guides through my year-long exploration of higher education during my ACE fellowship at Yale University in 2012–13. This would include my mentors, Peter Salovey, President of Yale University, and David Oxtoby, President of Pomona College, as well as many of my colleagues at Yale, Yale-NUS College, NUS and Pomona College. I am particularly grateful for the opportunity given to discuss STEM education with brilliant innovators such as Eric Mazur of Harvard University and Carl Wieman of Stanford University, who both visited Yale-NUS College in Singapore, and David Drew of the Claremont Graduate University, who has been instrumental in getting this book to print. Interviews with Jo Handelsman from Yale University, Kathy Takayama from HHMI, Eric Mazur of Harvard University, Anant Agarwal, CEO of edX, Richard Levin, former CEO of Coursera and President of Yale, Rob Lue from HarvardX, and Drew Endy from Stanford University added both depth to my research and a wonderful chance to meet and get to know some of the most innovative educational leaders in the world. I appreciate the contributions from Adam Nelson, who read and corrected an early draft of the first chapter. And of course, I am extremely grateful to my family—particularly my rocket-scientist wife Bidushi and my brilliant daughters Shanti and Asha—for their enduring love and support, and also for their perspectives in science and education.

Contents

About the Author

Prof. Bryan Edward Penprase is the Dean of Faculty for the Undergraduate Program at Soka University of America, in Aliso Viejo, CA. He also has been an American Council on Education Fellow at Yale University, Visiting Associate at the California Institute of Technology, and formerly was a Professor of Science at Yale-NUS College, and the Frank P. Brackett Professor at Pomona College, in Claremont, CA. As Dean of Faculty, Bryan directs the undergraduate program at Soka University of America, where he works to advance the innovative undergraduate liberal arts curriculum and is developing a new interdisciplinary concentration in Life Science.

Bryan received both a B.S. in Physics and an M.S. in Applied Physics from Stanford University in 1985, and a Ph.D. from the University of Chicago in Astronomy and Astrophysics in 1992. Prof. Penprase conducts research on the most luminous sources of radiation in the universe: quasars, supernovae, gamma-ray bursts, and merging neutron stars, in the context of his collaboration with astrophysicists at Caltech and worldwide on the "Zwicky Transient Facility" or ZTF. The research on such transients has discovered new types of supernovae, new gamma-ray bursts, and gravitational lensing sources. Bryan's research includes nearly all aspects of observational astrophysics, from photometric observations of nearby asteroids to spectroscopic studies of element formation in the Early Universe, using telescopes such as the Hubble Space Telescope and the Keck Telescope in Hawaii.

Bryan is the author of "The Power of Stars—How Celestial Observations Have Shaped Civilization," published by Springer, Inc., numerous book chapters on liberal arts education and STEM curriculum, and over 50 peer-reviewed articles, in the Astrophysical Journal, Astronomical Journal, and in Nature and Science. Bryan has served on numerous NSF and NASA review panels, including the Hubble Space Telescope Time Allocation Committee and the NASA/Keck Time Allocation Committee, and has participated in the external review of the Five College

Astronomy Program. His most recent research program is a collaboration with Caltech to develop the Zwicky Transient Facility (ZTF) and a Global Relay of Observatories known as GROWTH for studying gamma-ray bursts, new supernovae, and the electromagnetic counterparts of gravitational wave sources.

http://bryanpenprase.org

http://sites.soka.edu/faculty/bpenprase

Chapter 1
History of STEM in the USA

Abstract A review of the history of Higher Education in the US, with an emphasis on the role STEM education and diversity within in US higher education from Colonial times to the twenty-first century. The founding of the first universities in the US was motivated by providing religious training and later shifted toward science and engineering education as the nation began to grow and the education system included more diverse institutions. At each stage the growth of higher education produced new economic growth, and yet the expansion of the higher education system made slow progress in providing greater access to higher education for women and non-white students. In the nineteenth century the founding of liberal arts colleges, historically black colleges and Land Grant institutions began to provide more diverse curricula and provided access to women and African American students. In the twentieth century, US universities were massively scaled up, and increasing fractions of students received degrees in STEM subjects, with the growth of a STEM workforce which enabled social mobility through higher education. The concept of a meritocratic system of equal opportunity, underlies the American Dream and expectations for social mobility in the US has not provided equal results for students from all racial and ethnic groups, and especially in STEM fields, where the levels of enrollment, degree completion and persistence through graduate programs are all lower for non-white students and women. Bringing greater equity and inclusion within STEM fields is urgently needed, and the chapter reviews some key recommendations for helping develop more diversity in the STEM workforce across all levels of higher education.

Keywords Diversity · Inclusion · Social mobility · Education history · STEM pipeline · STEM equity

1.1 History of US Higher Education and Diversity in STEM

Ever larger numbers of our students, from all demographic, ethnic and socioeconomic groups need access to higher education, which has always played a central role in promoting social mobility and democracy in our country. The success of our economy and our higher education system requires that all of the most talented students have equal access to careers in Science, Technology, Engineering and Math

B. E. Penprase, *STEM Education for the 21st Century*,
https://doi.org/10.1007/978-3-030-41633-1_1

(or STEM), which often provide rewarding careers and a chance to develop future technologies. Historic inequities, and unequal access and funding for education for students of different economic and ethnic groups has unfortunately resulted in serious disparities between the demographics of STEM professionals compared to the larger US population. The increasing fraction of Latino/a and African American students in higher education is not reflected in faculty ranks, especially in STEM fields. Developing greater diversity in STEM will enable our scientists and engineers to embody the full diversity of the population, which will not only advance economic competitiveness but also assure greater equity and social justice.

As the former President of Yale University, A. Bartlett Giamatti, eloquently wrote in his 1988 book *A Free and Ordered Space*:

> The University today is very different from the one twenty-five years ago, or fifty or one hundred or two hundred and fifty years ago, and yet it is not different. It is still a constant conversation between young and old, between students, among faculty; between faculty and students; a conversation between past and present, a conversation the culture has with itself, on behalf of the country. The university lives through all its voices… Perhaps it is the sound of all those voices, over centuries overlapping, giving and taking, that is finally the music of civilization, the sound of human beings shaping and sharing, mooring ideals to reality, making the world, for all its pain, work. The university is the place where the seeds of speech first grow and where most of us first begin to find a voice. (Geiger 2015; Giamatti 1988)

Increasing access and success of all of our students in higher education and in STEM fields regardless of their race and socioeconomic class will enable them to partake in the "music of civilization," and to help "make the world work." As Giamatti said of Yale—"to be truly a national institution, playing its educational and civic role to its fullest, Yale's texture can never be less varied and many-grained than the fabric of America itself." This statement is one that applies equally to our entire system of STEM education.

In their influential work *The Shape of the River* (Bowen and Bok 1998), William Bowen and Derek Bok noted that "the growing diversity of American society and increasing interaction with outside cultures worldwide" made institutions that lacked diversity "increasingly anachronistic." They noted the benefits of having a diverse campus included "being able to understand how others think and function, to cope across racial divides, and to lead groups composed of diverse individuals" (Bowen and Bok 1998). The rapid pace of change in twenty-first century science in fields such as biotechnology, artificial intelligence, nanomaterials, and computer technologies, places an ever-higher premium among STEM professionals to regularly collaborate with diverse teams of researchers, and also will require us to develop all of our available talent within society.

1.2 Attitudes About Higher Education and the American Dream

An aspiration for equal access to higher education, at least across socioeconomic lines, has been part of the rhetoric and aspiration of the US throughout its history, despite terrible and persistent racial inequities. Slave-owner Thomas Jefferson described equal access to education across socioeconomic class as the means for an developing an "aristocracy of talent:"

> "instead of an aristocracy of wealth, of more harm and danger, than benefit, to society, to make an opening for the *aristocracy of virtue and talent*, which nature has wisely provided for the direction of the interests of society, and scattered with equal hand through all its conditions.... (Jefferson 1821)"

Even before the US existed as a nation, it enthusiastically built colleges. The first colleges in the US include Harvard (1636), Yale (1701), University of Pennsylvania (1740), Princeton (1746), Columbia (1754), Brown (1764), and Dartmouth (1769). This Colonial system of higher education was largely fueled by religious fervor and the need to train ministers. Harvard's charter, enacted in 1650, also expressed an early commitment to diversity, requiring that both Native American and English students be admitted to "conduce to the education of the English and Indian youth of this country in knowledge and godliness." Yale's charter from 1701 reflected an emphasis on both religion and science, so that "Youth may be instructed in the Arts and Sciences [and] through the blessing of Almighty God may be fitted for Public employment both in Church and Civil State." The College of Rhode Island (later Brown University) was founded in 1764 with religious diversity as a core principle, stating that, even if Baptists held a governing majority, "into this liberal and catholic institution shall never be admitted any religious tests, but on the contrary, all the members hereof shall forever enjoy full, free, absolute, and uninterrupted liberty of conscience" (Library of Congress 2010).

With instruction and "godliness" as a primary focus of these first US institutions, what little science and mathematics existed was based on the European medieval curriculum in liberal arts derived from the *trivium* and the *quadrivium*. The *trivium* comprised the subjects of Latin Grammar, Rhetoric, and Logic, which resulted in the degree of baccalaureate, and were considered the foundations for undergraduate education. Science entered into the medieval liberal arts through the *quadrivium*, and included studies of arithmetic, geometry, music and astronomy. These "seven liberal arts" were very slow to adapt to changing societal needs. Harvard College in the early eighteenth century offered the subjects of Rhetoric, Logic, Ethics, Politics, Arithmetic, and Geometry, all in Latin. Additional optional studies in Algebra, Astronomy, Physics, Metaphysics and Theology were available, however, for more advanced students (Harvard University 2017). Newtonian Physics did not become part of the standard curriculum within the United States until the mid-eighteenth century, and students were taught Aristotelean physics long after it had been disproven by experiment. Observational astronomy was only offered in the wealthiest institutions, like Harvard, that could afford to have a telescope. As described by historian

Edmund Morgan, "Students were not presented with opposing views and asked to choose between them. Instead they were expected to assimilate Aristotelian rhetoric, Ramist theology, Berkeleyan metaphysics, and diluted Newtonian Physics" (Geiger 2015).

1.3 Development of Science Education in the United States

With the arrival of the Industrial Revolution and the expansion of the US into vast new territories, new institutions and new curricula were developed that offered STEM subjects. At least for primary schools, the 1787 Northwest Ordinance codified the connection between education and the success of the new nation, and stated that "knowledge, being necessary to good government and the happiness of mankind, schools and the means of education shall forever be encouraged." By the early 1800's, US higher education institutions began to proliferate, but limited enrollment almost exclusively to white males. New institutions such as the University of North Carolina (1795) did away with requirements for Latin and Greek, and instead required courses focused on English literature and Science. New state universities in Georgia and South Carolina, founded by academics from Yale and Union College, built in science programs adapted from their previous institutions. US higher education in the early nineteenth century often provided "a nursery for virtue and science, the two brightest pillars of republican government" (Hollis 1951). As one Methodist document from the early nineteenth century put it, students "should strive by every lawful means to imbue his mind with every science which is intimately connected with the doctrine of salvation by Jesus Christ" (Duvall 1928).

Small private liberal arts colleges during this period were established with a "distinctly American" emphasis on small residential communities which in some cases included more diversity in students and curriculum. Williams College, founded by former Yale faculty in 1793, was followed by Middlebury College in 1800, and Amherst College in 1821, founded mostly by Williams faculty and students. After 1820, dozens more liberal arts colleges opened, at a rate of one every 8 months in the 1820's and doubling to 3 per year from 1830 to 1845, and then again doubling to 10 new colleges per year by 1860. By the end of the nineteenth century, over 150 liberal arts colleges had been founded, including several in the Midwest and Western United States, such as Oberlin (founded in 1833), Willamette College (founded in 1842), Santa Clara College (founded in 1851), Carleton College (founded in 1866), and Pomona College and Occidental College (both founded in 1887).

Several of the new US liberal arts colleges expanded the diversity of their students, who previously were exclusively white and male. In 1804, Middlebury offered an honorary master's degree to Lemuel Haynes, an African American Revolutionary war hero. In 1823, Alexander Lucius Twilight graduated from Amerst College, in 1826 Edward Jones graduated from Amherst College (the second African American to earn a US college degree), and in 1826 John Brown Russwurm graduated from Bowdoin College. In 1833 Oberlin College's first class of students included several African

American and female students. Some new institutions, such as Mount Holyoke College (1834), were dedicated entirely to education for women. The founder of Mount Holyoke, Mary Lyon, was herself a chemistry instructor, and required seven courses in science and mathematics for all of the women at Mt. Holyoke, including laboratory experiments and field work in geology and botany. Many of the new liberal arts colleges, such as Pomona College, Occidental Colleges and Oberlin College, were co-educational from their founding, and provided many additional opportunities for women to study STEM subjects.

Science and Engineering education developed slowly but steadily in the US in the first half of the nineteenth century. The Hollis Science Professorships at Harvard and Yale's Professor Silliman gave some US students exposure to modern scientific research in the first part of the century, even as the majority of the US faculty and students remained firmly focused on classical subjects. In 1817, the West Point Military Academy offered the only dedicated engineering education in the US, emphasizing Civil Engineering and helping develop the U.S. Army Corps of Engineers (Smithsonian 2019). In 1819, the American Literary, Scientific and Military Academy opened in Norwich, Vermont, and was joined in 1822 by the Gardiner Lyceum in Maine, which also offered a "practical" scientific curriculum. By 1824, the new Franklin Institute in Philadelphia and the Rensselaer School in Troy, New York, offered specialized degree programs in science. The Rensselear School developed the first specialized degree programs in science, with a bachelor's of natural science available in 1824 and civil engineering courses by 1835 (Geiger 2015). As early as 1815, Union College developed a new science curriculum and introduced its own civil engineering program in 1845.

Alexis de Tocqueville noted the increasing access to education in 1830's America, and its strikingly pragmatic approach:

> the effort made in this country to spread instruction is truly prodigious. The universal and sincere faith that they profess here in the efficaciousness of education seems to me one of the most remarkable features of America... Science is studied as if it were a job and only those branches are taken up which have a recognized and immediate usefulness. (Tocqueville 1954)

American higher education entered a new era with the passage of the Morrill Act of 1862. The Morrill Act established "Land Grant" universities based on the provision of 30,000 acres of federal land for each member of a state's congressional delegation. These lands could be used directly for new universities or sold to fund the construction of new institutions. Within a few decades, 69 "land grant" colleges were founded, among which are some of the best universities of the world. Early Land Grant institutions include Cornell (1865), MIT (1865), the University of California (1868), and the University of Wisconsin (1866). The language of the Morrill Act emphasized the importance of access to higher education from diverse populations that included "the industrial classes" among the student body:

> ...by each State which may take and claim the benefit of this act, to the endowment, support, and maintenance of at least one college where the leading object shall be, without excluding other scientific and classical studies, and including military tactics, to teach such branches

of learning as are related to agriculture and the mechanic arts, in such manner as the legislatures of the States may respectively prescribe, in order to promote the liberal and practical education of the industrial classes in the several pursuits and professions in life. (National Archives 2019)

After the passage of the Morrill Act, segregation and the exclusion of African–Americans from all-white campuses required the founding of colleges catering exclusively to African–American students. From 1865 to 1880, 39 African–American colleges and universities were founded, including Fisk University (1866), Atlanta University (1867) (which included W.E.B. DuBois, the first African American to receive a PhD from Harvard among its faculty), and Howard University (1867). Only 208 African American students had received a bachelor's degree throughout the entire history of the US up until 1876, but in the four years of 1876–1880 this number doubled, with 215 degrees conferred, and in the 1880's, 1336 bachelor's degrees were conferred to African Americans (Johnson 1938; Haynes 2006).

Even with the rapidly growing number of private universities, colleges, and new land grant universities, the US higher education system in the nineteenth and early twentieth centuries provided training for only a very small fraction of the population, who were overwhelmingly white males. In 1890 a second Morrill Act was passed, explicitly prohibiting racial discrimination in colleges receiving federal funds. Many states evaded this regulation, however, by providing separate institutions for African Americans. This resulted in 17 new land grand institutions for African Americans being founded in 1890, including Tuskegee University, adding to the growing number of institutions that today include 100 historically black colleges and universities in the US (National Academies 1995).

The number of bachelor degree-granting institutions grew steadily through the twentieth century and beyond, reaching over 5000 institutions of higher education within the US by 2015 (Fig. 1.1). With the increasing capacity of US higher education, the number and percentages of 23-year old students receiving bachelor's degrees continued to rise as well, but the benefits were not shared equally across gender, race and class. By the beginning of the twentieth century, only 2–3% of the US population

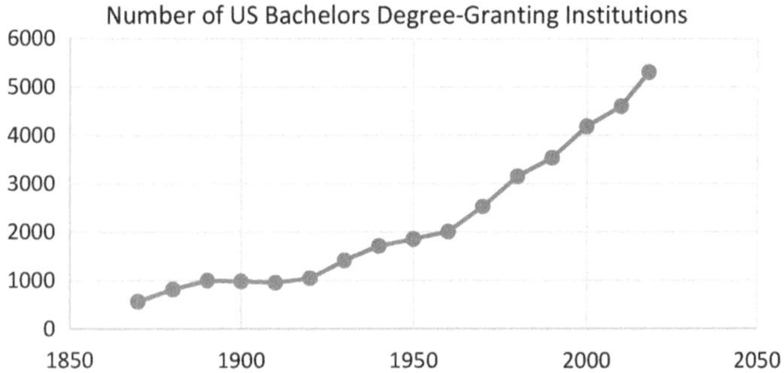

Fig. 1.1 Growth of number of bachelor degree granting US Institutions of Higher Education from 1870 to 2017, showing the growing capacity of US higher education (*Source* NCES 1993, 2017)

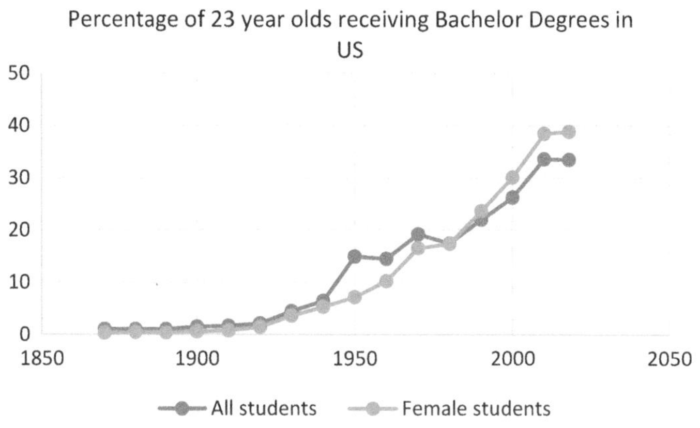

Fig. 1.2 Growth of the percentage of 23 year olds in the US receiving a bachelor's degree from 1950 to 2017 (*Source* NCES 1993, 2017)

possessed a college education, and this figure did not pass 25% of the population until after 1980 (Fig. 1.2), with far lower percentages for African–American and non-white students throughout the period, rising from 1% of the African–American students in 1940 to just over 10% in 1990 (NCES 1993).

1.4 STEM Education in the Twentieth Century

During the ninteenth and twentieth centuries, the US higher education infrastructure, industrial capacity, competitiveness and population increased at a rapid pace. Attendance at the American colleges and universities rose by an even faster amount—increasing by 400% in first half of the twentieth century, and accelerating even faster after World War II (Levine 1986). The spectacular expansion in US higher education during the twentieth century provided a corresponding explosion of scientific discoveries and rapid advances in technology, but the benefits of the expansion was not shared equally among racial and socioeconomic groups, and many campuses were effectively segregated by class and race. As David Levine writes in *The American College and the Culture of Aspiration*:

> while the ambitious of all backgrounds flocked to the colleges, most of the ambitious poor could not afford to attend. Furthermore, once there, though they all shared the same hopes, collegiate culture only occasionally brought young people from diverse socioeconomic backgrounds together in the same classes, activities, and fraternities. (Levine 1986)

American universities and colleges expanded their STEM offerings in all fields to keep up with the demand for highly trained specialists in the industrial economy, with STEM accounting for an ever-increasing fraction of degrees awarded. Throughout the second half of the twentieth century, US higher education continued its rise in both attendance and degree completion. The implementation of the GI bill after World War II accelerated the increase in the numbers and diversity of the student populations in the US. From 1945 to 1975, there was a six-fold increase in college degree attainment, rising from 5% of 23-year old students with a bachelor's degree in 1945 to over 30% by 1975, with an increasing fraction of those degrees in STEM fields. However systematic discrimination prevented African–American students from attending many universities, even after the 1954 *Brown versus Board of Education* decision which required desegregation in education, and only a few percent of African–American high school graduates were able to attend college, with a small minority of those students graduating in STEM fields.

By 1960, the percentage of bachelor's degrees awarded in STEM fields (including computer science, engineering, health sciences, life sciences, mathematics, physical sciences and psychology) accounted for 25% of degrees awarded. By 2010, the "STEM workforce" was growing by over 500,000 new degrees per year in STEM subjects. The diversity of this growing STEM workforce lagged well behind the population, however, with some improvement from 1988 to 2008, as shown in Fig. 1.3. During the period from 1988 to 1998, STEM enrollments by under-represented

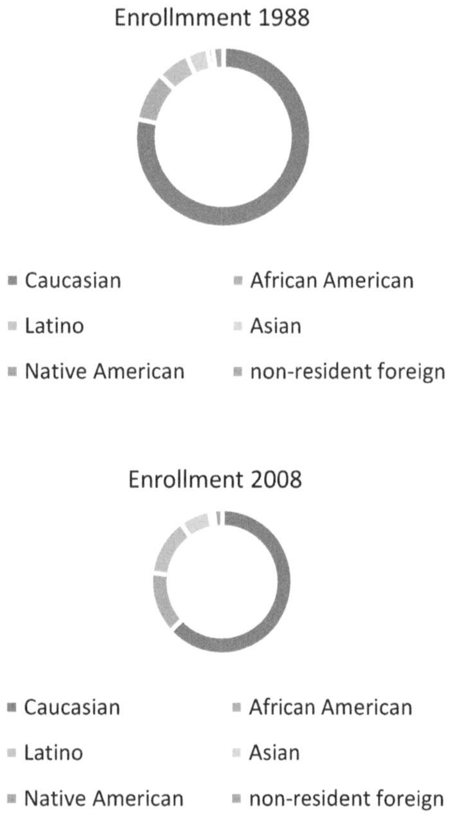

Fig. 1.3 Charts showing enrollments in STEM subjects by race and ethnicity, for the years 1988, and 2008, showing both the increasing diversity in STEM students within the US, but also the need for greater representation of non-white students to match US demographics

minority students nearly doubled, growing from 22% in 1988 to 37% in 1998 and rose to 42% in 2008, which held steady through 2015. Despite the growth in diversity in enrollment, degree completion for minority students in US institutions in STEM subjects lagged well behind white students, with only 30% of STEM degrees being received by non-white students in 2008, with a slight rise to 36% by 2015–16 (NCES 2017).

1.5 STEM Education and Social Mobility Within US Higher Education

Throughout the twentieth century, the growing demand for STEM expertise made STEM education a key component of social mobility within the US. This period was accompanied by long expanses of sustained economic growth, and a corresponding growth in US prowess in technical industries and innovation within a rapidly globalizing economy. Maintaining broad access to higher education, regardless of one's socioeconomic or ethnic background, is central in building "social mobility." Peter Salovey, President of Yale University, provides a personal example of how this sort of social mobility in the US enabled opportunities for him and his family to achieve their dreams:

> My father's parents—my grandparents—came to the United States by way of Warsaw and Jerusalem… Like many immigrants, my grandparents were poor in means but rich in culture and spirit. My grandmother worked as a bookkeeper, and after my grandfather lost his small store during the Depression, he took care of the elderly and infirm. But today I stand before you, a professor of psychology and president of one of the world's finest universities—the finest university. Today I stand before you, humbled, because I know that there are very few places in the world where, in two generations, a family can rise from modest means to this kind of role in life. (Salovey 2013)

Salovey connects this rise with the education that his father got at the Bronx High School of Science, and scholarships from City College of New York. These opportunities, coupled with hard work, are described by Salovey as the essential ingredients of the American Dream, which is advanced by policies to promote the broadest possible access to education:

> …this is the American Dream as well. Study hard, work hard, and you will be rewarded by social and economic mobility. Because of education, the ladder of opportunity was within reach for my father. This same vision, this same emphasis on education, has motivated progressive legislation from the Morrill Acts creating the land-grant colleges, to the GI Bill, to Title IV and the Federal Pell Grant program. These investments are worth making because college graduates develop broad and deep knowledge, sharp critical thinking and communication skills, and the ability to work collaboratively. (Salovey 2013)

Despite the rapid expansion of higher education within the US, social mobility has levelled off and now lags behind many European countries. One study observed that the US leads among all of the OECD countries in the influence of parental background on student achievement in secondary school, a factor that is anti-correlated with social mobility (OECD 2010). This same study showed that mobility in wages and earnings across generations was lower in the United States than in countries like Australia, Canada and the Nordic countries. Another 2006 study concludes that "mobility is lower in the U.S. than in the U.K., where it is lower again compared to the Nordic countries." The study further notes that there is a "particularly high likelihood that sons of the poorest fathers will remain in the lowest earnings quartile," which the study suggests could be a new definition of "American exceptionalism" (Jantti et al. 2006). More recent studies have shown further decreases in social mobility and increased income inequality. For example, the US currently is ranked 9th from the bottom among 44 OECD countries for income inequality, as measured by Gini score (OECD 2018). In addition to low social mobility, the US is also under-performing in producing STEM graduates compared to our economic competitors—who also in many cases are also economic and scientific partners. For example, in 2017, China awarded 4.7-million bachelors degrees in STEM subjects, while India produced 2.6-million graduates, compared to just 568,000 new STEM degree holders in the US (McCarthy 2017).

The Pew Charitable Trust's *Economic Mobility Project* noted how limited social mobility, which correlates with race, locks in historical inequities in society. The study noted the shocking conclusion that "only 4% of those raised in the bottom quintile make it all the way to the top as adults, confirming that the 'rags-to-riches' story is more often found in Hollywood than in reality" (Pew Charitable Trust 2016). The Pew study noted that 66% of those raised in the bottom quintile of the wealth distribution remain in the bottom two quintiles as adults, while 66% of those raised in the top quintile remain in the top two quintiles. African Americans were far less likely to begin in the top two quintiles (only 9% as opposed to 46% for whites), and experience an 11% lower social mobility rate compared with whites in the middle quintile, and a 23% lower social mobility rate compared to whites in the second to lowest income quintile.

1.6 Diversity Within US Campuses and in STEM Fields

The disparity of social mobility within the US arises largely from persistent disparities in access and retention in higher education based on racial and gender lines, which are only worse in STEM subjects. At each level of our higher education system, these disparities are amplified, as students from under-represented groups have larger dropout rates from college, and lower percentages advancing past undergraduate and graduate degrees. While 33% of white and 42% of Asian American students intending majors in science complete their degree in five years, only 18.4% of African American and 22.1% of Latino students complete their degrees in the same period. Only 6% of the US student population receives degrees in STEM fields by age 24, compared to 2.7% of African American and 2.2% of Latino students (National Academies 2011). These disparities have caused the STEM workforce to be remarkably different from the demographics of the US population, as shown in Fig. 1.4.

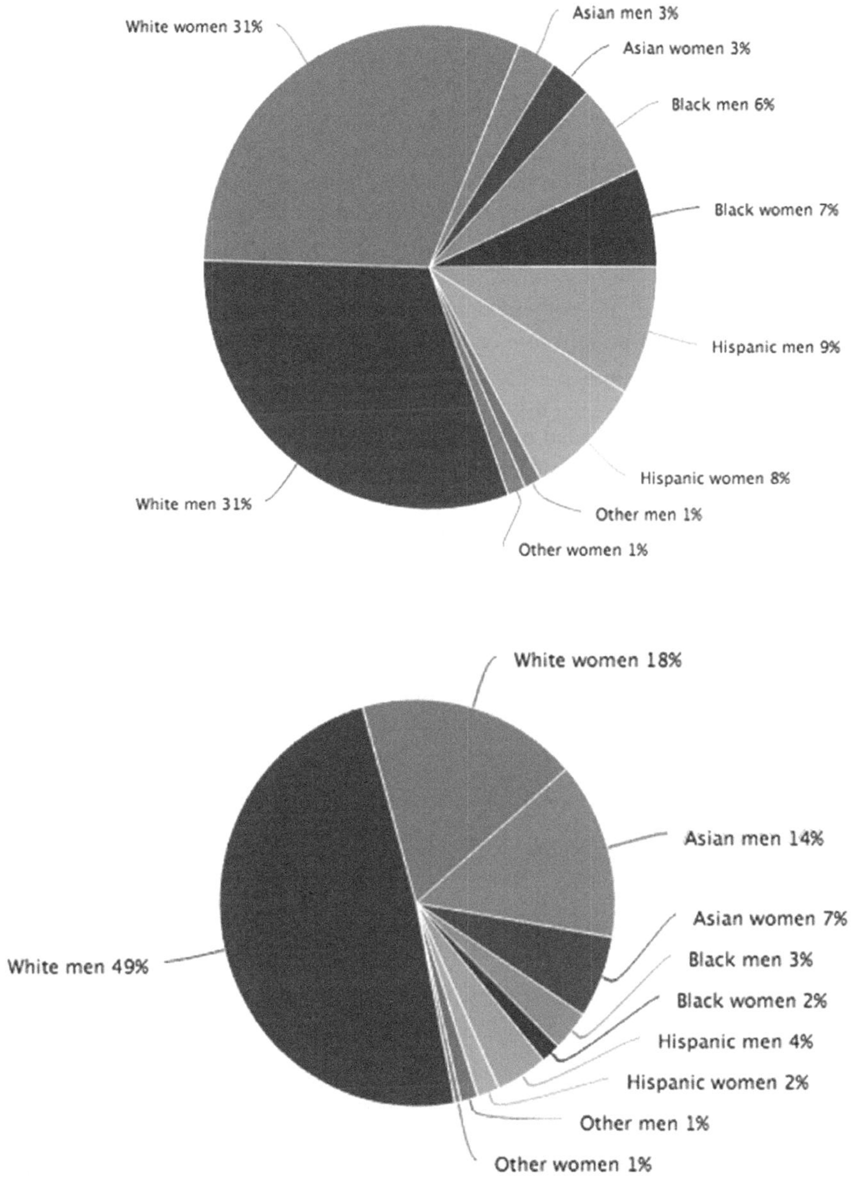

Fig. 1.4 Diversity of the overall US population (top) compared to the Science and Engineering workforce (bottom), based on 2014 statistics compiled by the National Science Foundation (NSF 2017a, b)

These disparities are only worse at higher levels of the "STEM pipeline." The Council of Graduate Schools report, *Graduate Enrollment and Degrees: 2002 to 2012* reported evidence of increased attrition for women and under-represented minorities at every level of the educational "pipeline". In one recent study URM students accounted for 17% of STEM bachelor's degrees (a factor of 2 less than their representation in the general population), but only received 5.4% of the awarded STEM doctorates (NSF 2017a).

As each of the levels of our STEM education system becomes less diverse, the pool of applicants from URM groups steadily decreases, resulting in a tiny number of candidates in many fields. This trend is shown graphically in Fig. 1.5, which shows the dramatic reduction of the fraction of URM students in STEM at each stage from high-school graduation to graduate school to PhD completion. As an example, in 2011 there were 288 PhD's awarded in astronomy and astrophysics, but only 3 went to African Americans. As another example, only 15 PhD's in Physics were awarded to African Americans in 2011 (out of 1777 total), and only 17 in Economics (out of 1124 total)! By 2018, these numbers have not improved significantly, with only 16 PhD's in physics awarded to African–Americans between 2014 and 2016, and only 6% of Physics PhD's awarded to URM graduate students during this period (AIP 2019). The small percentage of URM "survivors" of the STEM pipeline, typically only 2–4% in most fields (Fig. 1.6), make diversifying the faculty in STEM very challenging (National Academies 2011). The lack of diversity within faculty ranks also limits the diversity of role models and mentors for incoming students, further contributing to the disparity in outcomes in STEM across ethnic lines.

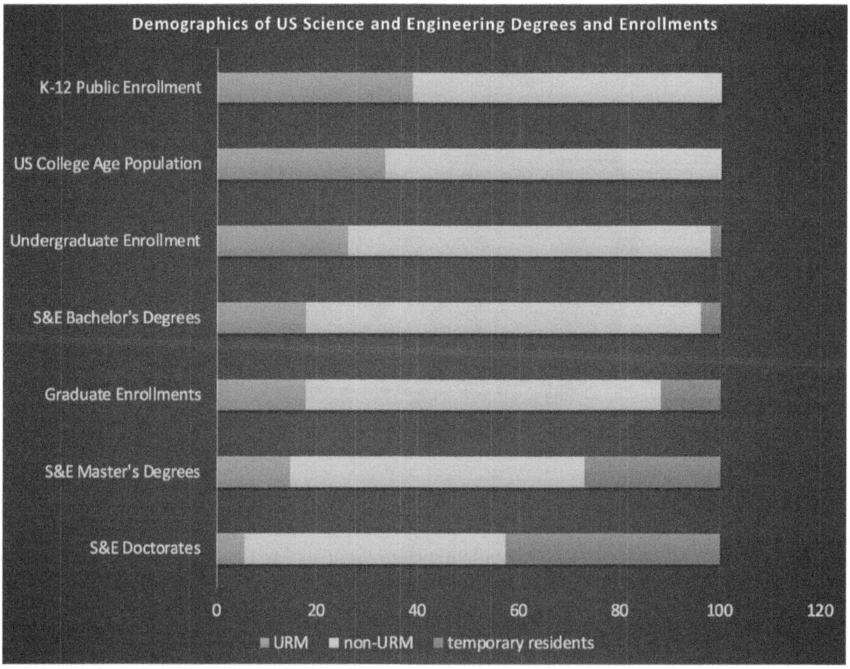

Fig. 1.5 Percentage of Underrepresented Minority (URM) population (grey) as a function of educational level, with K-12 enrollment at top and STEM PhD recipients at bottom. The graph shows a steady decline in the representation of URM from high school through Doctorate programs (NSF 2018)

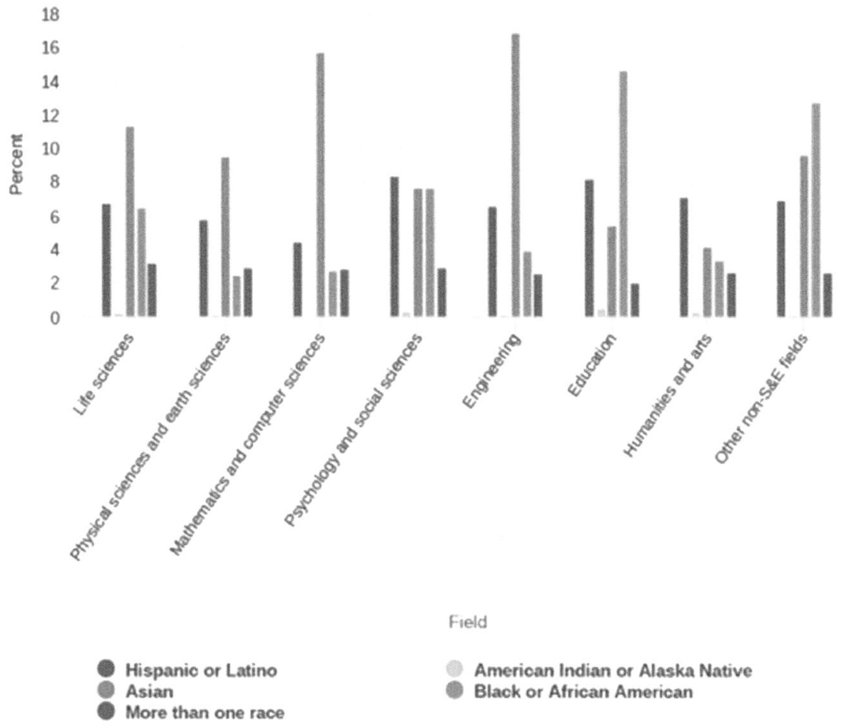

Fig. 1.6 Percentage of doctorates awarded to different demographic groups, based on field of study (NSF 2017b)

What is apparent from the serious lack of representation of under-represented groups across all levels of the STEM educational pipeline is that urgent attention needs to be paid to providing greater equity within STEM education in the US. The American Dream, with its narrative of meritocracy and social mobility, presumes a level "playing field" in higher education where students can compete equally for opportunities. However, this narrative neglects the historical legacy of systematic discrimination and injustice in US higher education that has been documented in the earlier part of this chapter.

The effects of historic inequities combine with widespread bias that still exists throughout academia. Multiple studies have shown that implicit bias plays a key role in disadvantaging women and under-represented groups in hiring and promotion and within classrooms (Banaji and Greenwald 2013). New instruments, such as the online Race IAT test (Project Implicit 2019) can allow faculty to measure their own implicit bias levels and begin to counter bias in their teaching, and to begin to implement more inclusive pedagogy that enables greater participation from all students, and fosters a more supportive environment within classrooms. By accounting for the effects of implicit bias, search and hiring committees can help overcome some of the barriers preventing more diversity in STEM fields. This greater emphasis on diversity and inclusion across universities and colleges will increase the retention of URM students and women in STEM fields, which over time can allow PhD programs and faculty ranks to draw from the full spectrum of talent within the US population and assure equity within US STEM education.

1.7 Conclusion—the Vital Need for Diversity, Equity and Inclusion

The necessity to diversify our faculty ranks and our student enrollments in STEM to fully reflect the talents and aspirations of our nation is vital—and yet diversity is only part of the process toward equity in STEM. Bringing systematic changes across our entire higher education system to monitor and enable students to succeed through diverse pathways to graduation using 2 and 4-year institutions, removing "weed-out" cultures within science departments, reforming science education to be more inclusive, and providing more elements of co-curricular support to assure success of all students is needed (Malcolm and Feder 2016). This emphasis on success for all students promotes equity, which is a more proactive and just outcome than the meritocratic notion of equal opportunity, which ignores the effects of systemic injustice and bias that create barriers to success for too many students. Developing equity in STEM requires an inclusive environment that provides improved mentoring, additional academic support for students, and an institutional commitment to make all groups on campus feel valued and welcome.

Diversity adds to the "music of civilization" that is produced within universities and colleges as each voice adds to the conversation on campus. Like any conversation, all voices need to be heard—both from the majority and minority, and both from traditionally powerful and traditionally marginalized groups. And it is only with the contributions from all of our people, will the potential of our nation to advance in science and technology be fully realized, and will the fields within STEM promote equity and justice for our society.

References

AIP. (2019). *Race and ethnicity of physics*. PhD, Classes of 2014 through 2016. Retrieved January 16, 2019, from http://aip.org/statistics.

Banaji, M. R., & Greenwald, A. G. (2013). *Blindspot*. New York: Delacorte Press.

Bowen, W. G., & Bok, D. C. (1998). *The shape of the river: Long-term consequences of considering race in College and University Admissions*. Princeton, NJ: Princeton University Press.

de Tocqueville, A. (1954). *Democracy in America*. New York: Vintage Books.

Duvall, S. M. (1928). *The methodist episcopal church and education up to 1869*. New York: Teachers College, Columbia University.

Geiger, R. L. (2015). *The history of American higher education: Learning and culture from the founding to world war Ii*. Princeton: Princeton University Press.

Giamatti, A. B. (1988). *A free and ordered space: The real world of the university*. New York: W.W. Norton.
Harvard University. (2017). *17th/18th century curriculum overview*. Retrieved January 14, 2019, from https://guides.library.harvard.edu/c.php?g=425353&p=2906742.
Haynes, B. (2006). *Black undergraduates in higher education: An historical perspective* (Vol. 17, No. 2). Metropolitan Universities. Retrieved December 30, 2019, from https://journals.iupui.edu/index.php/muj/article/view/20252.
Hollis, D. W. (1951). *University of South Carolina*. Columbia: University of South Carolina Press.
Jantti, M., et al. (2006). *American exceptionalism in a new light: A comparison of intergenerational earnings mobility in the nordic countries, the United Kingdom, and the United States*. IZA Discussion Paper No. 1938. Retreived January 16, 2019, from http://ftp.iza.org/dp1938.pdf.
Jefferson, T. (1821). *Autobiography*. University of Chicago. Retrieved January 14, 2019, from http://press-pubs.uchicago.edu/founders/print_documents/v1ch15s20.html.
Johnson, C. (1938). *The Negro College Graduate*. Chapel Hill, NC: University of North Carolina Press.
Levine, D. O. (1986). *The American College and the culture of aspiration, 1915–1940*. Ithaca: Cornell University Press.
Library of Congress. (2010). *Primary documents in American history—Morill Act*. Retreived January 14, 2019, from https://www.loc.gov/rr/program/bib/ourdocs/morrill.html.
Malcolm, S., & Feder, M. (2016). *Barriers and opportunities for 2-year and 4-year STEM degrees: Systemic change to support students' diverse pathways*. National Academies Press. Retrieved December 30, 2019, from http://nap.edu/21739.
McCarthy, N. (2017, February 2). The countries with the most STEM graduates. *Forbes Magazine*. Retrieved January 14, 2019, from https://www.forbes.com/sites/niallmccarthy/2017/02/02/the-countries-with-the-most-stem-graduates-infographic/#683d3683268a.
National Academies. (1995). *Colleges of agriculture at the land grant universities: A profile, Chapter 1*. National Academies Press. Retrieved December 30, 2019, from https://www.nap.edu/catalog/4980/colleges-of-agriculture-at-the-land-grant-universities-a-profile.
National Academies. (2011). *Expanding underrepresented minority participation*. Report from the Committee on Underrepresented Groups and the Expansion of the Science and Engineering Workforce Pipeline (US). Washington, DC: National Academies Press.
National Archives. (2019). *Morril Act (1862)*. Retrieved Januray 14, 2019, from https://www.ourdocuments.gov/doc.php?flash=true&doc=33.
NCES. (1993). *120 years of American education: A statistical portrait*. Retrieved January 14, 2019, from https://nces.ed.gov/pubs93/93442.pdf.
NCES. (2017). *Digest of educational statistics: Table 318.45*. Retrieved January 14, 2019, from https://nces.ed.gov/programs/digest/d17/tables/dt17_318.45.asp.
NSF. (2017a). *Women, minorities, and persons with disabilities in science and engineering*. Retrieved January 14, 2019, from https://nsf.gov/statistics/2017/nsf17310/digest/introduction/.
NSF. (2017b). *Doctorate recipients from U.S. Universities*. Retrieved December 29, 2019, from https://ncses.nsf.gov/pubs/nsf19301/.
NSF. (2018). *National Science Board—Science and Engineering indicators 2018*. Retrieved December 29, 2019, from https://nsf.gov/statistics/2018/nsb20181/data/appendix?achapter561.
OECD. (2010). *Economic policy reforms: Going for growth 2010*. Retrieved January 14, 2019, from http://www.oecd.org/eco/growth/economicpolicyreformsgoingforgrowth2010.htm.
OECD. (2018). *Economic policy reforms 2018—Going for growth interim report*. Retrieved January 14, 2019, from https://www.oecd-ilibrary.org/economics/economic-policy-reforms-2018_growth-2018-en.
Pew Charitable Trust. (2016). *Financial security and mobility*. Retrieved January 14, 2019, from https://www.pewtrusts.org/en/projects/financial-security-and-mobility.
Project Implicit. (2019). *IAT website*. Retrieved December 30, 2019, from http://implicit.harvard.edu/.

Salovey, P. (2013, August 24). Yale and the American dream. *Yale News*. Retrieved January 14, 2019, from https://news.yale.edu/2013/08/24/yale-and-american-dream-president-peter-salovey.
Smithsonian. (2019). *West point in the making of America*. Retrieved January 14, 2019, from http://americanhistory.si.edu/westpoint/history_2.html.

Chapter 2
Active and Peer-Based Learning

Abstract The chapter reviews the history of how STEM educators were able to document more effective learning through assessments that provided quantitative measures of learning in active-learning environments. Key innovators within the field of active learning, peer-based learning and scientific teaching are profiled in detail, starting with leading innovators in the field of physics, were much of the best data validating active learning was first acquired. Eric Mazur's peer learning techniques, which enable students to discuss problems within large classes and therby "construct knowledge" are described detail. The history of Carl Wieman's Science Education Initiative is described, with details about his transition from Nobel Prize winning physics researcher into an internationally recognized leader in STEM education. The transformation of physics and astrophysics with active learning and research-based curricula is described with numerous examples for making more exciting and engaging classes through active learning. The development of active learning in Biology is described, including the work in BioQuest, SEAPhages and a new initiative from Jo Handelsman known as TinyEarth. These research-based courses and approaches to active learning and course design are described in detail to give examples of how 21st century research can be brought into the classroom to bring students in active engagement with cutting-edge science.

Keywords Active learning · Peer learning · Learning taxonomy · Scientific teaching · Research-based learning · Project-based learning

2.1 Students Constructing Knowledge

The "Copernican Revolution" is a classic example of a "paradigm shift" as described by Thomas Kuhn in his work *The Structure of Scientific Revolutions* (Kuhn 1996). In the Copernican example, astronomers had to adopt a new model with the sun at the center of our solar system, and the earth was demoted to just one of several planets orbiting the new sun. A similar shift is going on in science teaching—in this case the shift is from a "professor-centered" model of teaching based in lecture format, toward a "student-centered" model of active learning. The mode of "active learning" has replaced lectures with hands-on activities, concept tests, peer learning,

B. E. Penprase, *STEM Education for the 21st Century*,
https://doi.org/10.1007/978-3-030-41633-1_2

and other types of teaching that have been demonstrated to increase the effectiveness of classroom time. This discovery has revolutionized how science is taught across the country, and new courses, programs and departments have been built on these new techniques of teaching.

By using new course design principles such as "Scientific Teaching," "Backward Course Design," or "Constructive Alignment," instructors can create courses that are well-designed mechanisms for enabling learning. Combining classroom activities that are aligned with learning outcomes that are assessable and based on practice, students and instructors can more accurately discern progress toward deeper understanding within courses. These course design principles when coupled with "active pedagogy" have been shown by many studies to provide deeper engagement in class, and more lasting learning. The growing awareness of the need for this type of course design and teaching has swept across the STEM education community, which while resistant to change, is slowly adapting new courses and educational strategies.

How did all of these advances take place? And what enabled professors to voluntarily change their ways? The story is an interesting blend of academic and disciplinary cultures, which largely arose from the community of scientists within physics and astronomy—my own disciplinary home. To provide an example of the development of active learning, I describe my own Copernican moment when I realized there is much more to teaching than lecturing, and that teaching, like science research, requires extensive collaboration. This means examining the developments outside of one's own classroom, one's own subject and one's own institution.

My own odyssey into active learning began at my former home institution, Pomona College, where I taught physics and astronomy for over 20 years, and was followed by journeys across the US and throughout Asia. I continued to meet educators who described a more active kind of classroom, and who had abandoned the traditional lecture—with results that included more student engagement, better learning and higher teaching evaluations. Overcoming the fear that many instructors feel in stepping outside of the traditional bounds of lecture-based science instruction was something that took nearly two decades in my case, and required collaboration with others who were able to share their stories and specific techniques for implementing this kind of teaching. From this journey, I learned that it is entirely possible to collaborate on course development, assessment, and learning outcomes, just as scientists regularly collaborate in their disciplinary research. I wanted to share with others how one can break out of less effective modes of instruction in STEM education and to be liberated from the "tyranny of own's own experience" which is necessarily limited to small numbers of students, courses and outcomes. The exploration of this process and journey is described below.

2.2 The Force Concept Inventory and Active Learning in Physics

At Pomona College in Claremont CA, where I taught for 20 years, my colleague Tom Moore has been a leader in physics education for decades with his "Six Ideas That Shaped Physics" curriculum, and his approach of having students discuss and work on problems within class. Years before clickers were invented, Tom was gathering feedback in class with cards showing multiple choice answers (A, B, C, D or E), or small whiteboards for equations. Tom's efforts created a widely adopted curriculum for introductory physics that not only features advanced topics usually only seen in later courses (quantum mechanics, relativity, and statistical mechanics), but also uses most of class time for students to work on problems instead of listening to lectures. When Tom began doing this work in the early 1990's it was revolutionary, but new research has shown that this approach results in large learning gains over a traditional course. Tom's own results confirm the effectiveness of the class, as his "problem based learning" teaching style gives larger gains in the widely used physics test known as the Force Concept Inventory than traditional classes (Moore 2019).

The Mechanics Diagnostic Test (Halloun and Hestenes 1985), which evolved into the "Force Concept Inventory" or FCI (Hestenes et al. 1992) provides a powerful tool for measuring student learning in basic physics and the growth of student insight into underlying physical concepts. It was only from having this widely accepted quantitative instrument for measuring student learning that physicists (who naturally love quantitative data) were convinced of the power of active learning. The FCI has been widely used in all forms of physics courses—big classes, small classes, interactive classes, studio physics classes, lab classes—enabling comparisons in learning gains using this same test. It also helps that the physics community, perhaps more than other STEM disciplines, has traditionally shared a common "canon" of introductory physics that begins with Newtonian mechanics—with the familiar array of sliding blocks, pulleys on weights, cannons shooting projectiles and colliding frictionless carts. The combination of a widely accepted instrument for measuring student learning, and a largely interchangeable curriculum in introductory physics allowed for physics to take an early lead in experimenting with the effects of active learning.

These experiments began in the early 1990s and continued on for decades within a wide range of institutions, with varying class sizes and demographics. The use of the FCI allowed the rare chance to quantify the "gain" students experience from different teaching approaches and to isolate those gains within different student populations. One example of the kind of data produced within the FCI is shown in Fig. 2.1, which includes a plot of FCI data from a variety of high school, college, and university settings using both "Traditional" lecturing and the more modern active learning or "Interactive Engagement" techniques. The active learning approach produced much larger gains in FCI test scores, with values of 0.62 ± 0.10, compared to a typical gain of 0.23 ± 0.04 in a traditional course (Hake 2005). These gains were documented in numerous experiments in classrooms across the country, where instructors agreed to administer the FCI before and after the course. Courses at the same institution, with

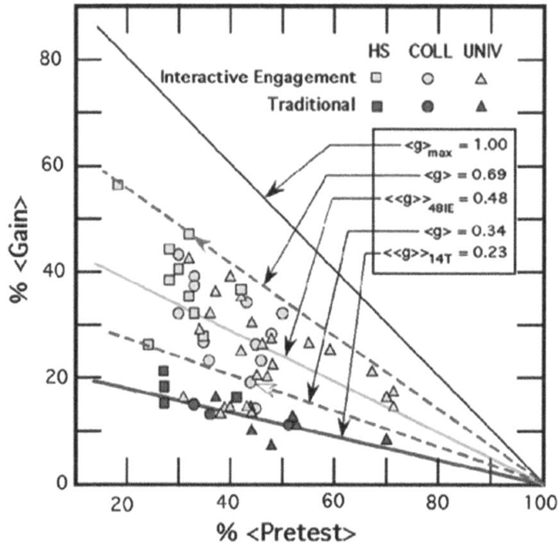

Fig. 2.1 Percentage gains on Force Concept Inventory tests, showing larger values of gains from active learning (green) than in "traditional" lecture courses (red) in high school, college and university settings (Hake 2002)

the same instructor teaching the same basic demographics of students showed significantly larger gains with active-learning compared with lecture-based approaches. The rigorous and quantitative analysis of FCI results proved very persuasive and allowed for the physics education community to validate the effectiveness of active learning within their classrooms within a single semester.

2.3 Eric Mazur and the Development of Peer Instruction

A key component of active teaching includes having students reason with each other in class about actual physics problems in a technique known as Peer Instruction. Eric Mazur from Harvard is one main pioneers of this technique, and has refined the techniques for posing the questions and selecting student groupings to maximize learning. Peer Instruction enables large classes to break into small groups of 3–4 students working on problems during the class and reporting their results back to the instructor. One key tool in this approach for larger classes is the "clicker"—infrared remotes attached to a control unit—that can instantly record and display the answers coming from the class. The clicker software and hardware has matured to a point where they are very easy to use, and now classes of several hundred can be fully as interactive as Tom Moore's small Pomona classes using the combination of peer learning groups and clicker technology. With new software such as PollEV and

Socrative, clickers can be replaced by student smart phones, to quickly give both multiple choice feedback and free response answers in class.

By monitoring the performance of particular students on in-class questions, Eric Mazur can now identify the higher performing students, and dynamically assign groups to mix skill levels and enable students to teach each other during the class and assure that all the group members understand the exercises. An extensive global community of instructors has been developed through an international Peer Instruction Network (Mazur and Schell 2019), which has over 2000 members across the world who are able to share ideal peer learning exercises that optimize the experience and learning in physics classes.

Eric began his odyssey into peer instruction and active learning back in 1990, when he realized that despite giving "clear and polished lectures and demonstrations" and receiving strong student evaluations, his students were learning "next to nothing" (Lambert 2012). This realization came from applying the Force Concept Inventory, which showed him that his lectures were not able to help students transition from manipulating equations to deeply understanding and applying fundamental concepts in physics. He began to experiment with giving students in-class opportunities to formulate answers to problems based on their own thinking, followed by discussions with their neighbors. The discussions allowed students to both articulate the conceptual basis behind their answers, and to reconcile any discrepancies. The technique was written up into his book *Peer Instruction: A User's Manual* (Mazur 1996) and has inspired hundreds of instructors to adopt the technique of "think-pair-share" within their classes. Research on ten years of data within peer instruction and active learning environments in physics (Crouch and Mazur 2001) has validated the approach, and demonstrated substantial gains in both the Force Concept Inventory and in traditional problem solving within active and peer learning environments. The more interactive learning environment has also shown even greater increases in learning among female physics students, helping to equalize the performance of women and men in physics courses. The peer instruction techniques can be used in any STEM course, or even humanities courses, and can be summarized into a simple set of principles. The implementation includes the following steps (Mazur 2009):

(1) Give a brief lecture (10–15 min) leading to a challenging conceptual question.
(2) Allow students to first think for 1–2 min about the answer.
(3) Have students individually show their answers (can be done with hands or fingers, cards, or clickers).
(4) If 35–70% have correct answer, ask students to find another with a different answer to discuss in pairs or small groups.
(5) After discussion have a re-vote, discuss the correct answer briefly, and move on if more than 70% have the correct answer.

This process can also be diagrammed as a flow chart in Fig. 2.2 to help show visually how an instructor would conduct a class, with branch points between more explanation and more discussion based on the responses from students.

Eric Mazur has helped STEM education shift toward active learning and away from lectures, a format that was developed "close to 1000 years ago." The new form

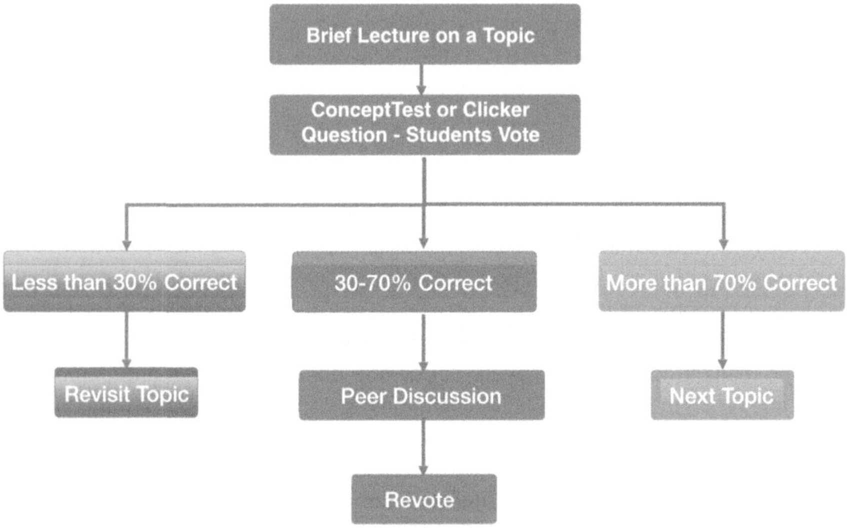

Fig. 2.2 Flow chart schematic of Eric Mazur's Peer Learning technique—showing sequence for using peer learning within a classroom, and decisions for whether to provide more information or a discussion based on student responses

of active learning enables the instructor to shift their role from "the source and the deliverer of knowledge" to "more that of a facilitator, of a coach." Eric believes there is much still to be done, particularly in the area of grading and assessment, which in his opinion often is producing shallow "surface learning" and in some cases is even "the silent killer of educational innovation" or "the silent killer of learning." The shifts in both teaching techniques and assessment will promote "more meaningful approaches to helping students learn" (Mazur 2013). Eric Mazur and Julie Schell have created a Peer Instruction Blog that is continuously updated with more tips and guidance on how to structure and optimize active and peer learning in classes from all disciplines (Mazur and Schell 2019).

2.4 Carl Wieman and Expert Thinking

One major force who has been instrumental in leading STEM education toward active learning in class is Carl Wieman, the 2001 Nobel Laureate in physics, director of science education initiatives at the University of British Columbia and the University of Colorado, and most recently at Stanford University. His journey as a teacher is described in a very thoughtful article in *Change* magazine. Carl describes what he calls the "learning puzzle":

> When I first taught physics as a young assistant professor, I used the approach that is all too common when someone is called upon to teach something. First I thought very hard about the topic and got it clear in my own mind. Then I explained it to my students so that they would understand it with the same clarity I had. At least that was the theory. But I am a devout believer in the experimental method, so I always measure results. And whenever I made any serious attempt to determine what my students were learning, it was clear that this approach just didn't work. (Wieman 2007a, b)

Carl reports that his shift in thinking about teaching physics was inspired "only after becoming sensitive to the failings of undergraduate instruction through observing new grad students in lab." He began to study the problem by undertaking a careful study of undergraduate understanding of physics, using both the FCI and focus group discussions. He observed that most students achieved very little learning during undergraduate courses, and only moved from "novice" to "expert" when they were advanced PhD students working on experiments. Carl realized that the active learning from these experiments integrated disconnected bits of information of physics into a "coherent structure of concepts that describe nature and that have been established by experiment." The trick is to create this same type of experience for students in the undergraduate classroom, instead of waiting 7 years for them to become advanced graduate students!

Carl identified from the research a series of research-validated steps necessary to improve STEM teaching. The first step is to "reduce cognitive load" in students, who suffer when their short-term memory is overloaded by a content-heavy lecture. Slowing down, using well-designed and concise figures and less jargon, all help in making sure that students learn. The second step is to "address prior knowledge and beliefs" that students bring to class. Instead of thinking that students are blank slates, their preconceived ideas and misconceptions actively help learning or resist concepts that contradict this "private universe." Third, instructors should motivate a topic by connecting the topic to things students already are familiar with, and with applications in the real world. Fourth, within the classroom environment, instructors should "stimulate and guide thinking" by designing activities to ensure students practice the desired cognitive processes, with frequent assessments to help them improve. And finally, Carl advocates using technology to extend the capabilities of the instructor in making the classroom more active—clickers and interactive computer simulations are two examples.

Adopting these new approaches can be difficult for professors, however, as Carl describes in his article "The Curse of Knowledge—or Why Intuition about Teaching Often Fails." Many professors who have advanced knowledge have a very difficult time seeing the subject from the perspective of someone who does not. Brain scans show that the cognitive patterns within "experts" are substantially different from the "novices." In Carl's words:

> This fundamental difference between the novice and expert brain explains many of the findings reported by those who study student learning of physics. Students can think about a topic in ways quite unimagined by the instructor, and so a lesson that is very carefully thought out and is beautifully clear and logical to experts may be interpreted totally differently (and

> incorrectly) by the student. This "curse of knowledge" means is that it is dangerous, and often profoundly incorrect to think about student learning based on what appears best to faculty members, as opposed to what has been verified with students. (Wieman 2007b)

Carl Wieman's articulate case for active learning to foster "expert thinking," has inspired a proliferation of new ways to teach physics and astronomy over the past decade. However, his research does not stop at the level of the classroom. Carl has thought deeply about how curriculum, departments, and even entire universities can be transformed to offer better teaching and learning for students. The result of this effort has been the Carl Wieman Science Education Initiative (CWSEI), and he has written up some of the lessons learned from this initiative in a recent book (Wieman 2017).

During a visit to Stanford University in 2016, I asked Carl about how places like Stanford and Yale-NUS College can best design a new curriculum. Carl noted that often there is a "disconnect" between curriculum design and pedagogy—leading to many problems. He indicated that it is not enough to design the content of the course (which is the more common and easy discussion) but that the curriculum design and the learning spaces need to be developed around active and engaged pedagogy. Carl noted that "no data shows that changing curriculum causes substantial changes. But we do have data on large effects from changing how things are taught. The teaching methods are what primarily determines how well students learn."

2.5 The Carl Wieman Science Education Initiative

The CWSEI was a well-structured program based on organizational change, which included a number of elements that have enabled several research universities to improve teaching with both resources and incentives for faculty to improve their teaching skills. The initiative goes beyond teaching evaluations to provide more thoughtful ways of assessing how people teach, and then use these assessments as a way of both improving teaching at the university, and incentivising faculty to upgrade their teaching practice. One of the main tools is the Teaching Practices Inventory or TPI (Wieman and Gilbert 2014) which helps faculty report on their pedagogical approaches while also helping institutions assess and improve the institutional teaching culture. The TPI is short survey to help faculty assess how they are teaching in an objective way which gives universities data to document the extent of effective teaching practices.

Carl argues that the TPI evaluates teaching in ways that are similar to how universities evaluate research on the basis of grants and publications. The Teaching Practice Inventory enables a quantitative measure that is a proxy for teaching quality, as one piece of the teaching review (Wieman 2015). Carl described two of his big experiments in institutional change—at the University of British Columbia and at University of Colorado, Boulder. He pointed out that at UBC 75% of faculty changed how they taught, which to his knowledge is the biggest change ever, amounting to 15,000

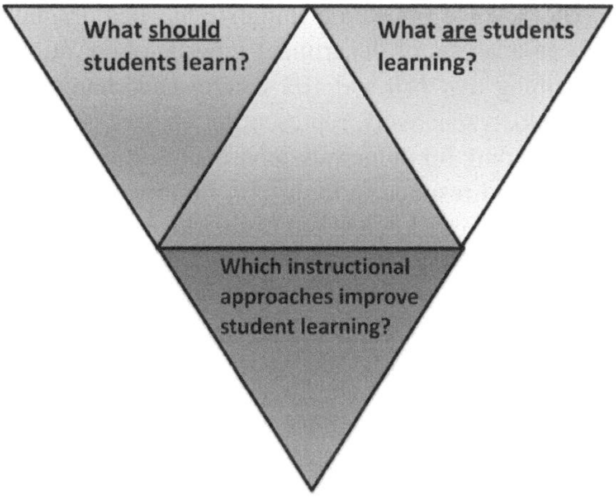

Fig. 2.3 The triangle of learning which needs to be addressed within teaching, based on the Carl Wieman Science Education Initiative (CWSEI)

credit hours per year. The training of faculty and postdocs all emphasises what Carl sometimes calls "the holy trinity"—knowing what students *should learn* from learning objectives that are assessable, being aware of *which instructional approaches improve learning*, and *assessing student learning* (and teaching) thoroughly, and adapting and adjusting approaches based on this information. The 'trinity' is illustrated in a simple diagram below that emphasizes the interconnections between these elements (Fig. 2.3).

Carl noted that any kind of educational reform effort at a university has to make use of the most important unit in the research organization—the department—which has shared expertise and culture that needs to be leveraged and respected. The trick to Carl's reform effort at UBC, Colorado and now Stanford is the use of "Science Educational Specialists" in departments. The Specialists typically have PhDs in the field, and are interested in a career in STEM education. Science Education Specialists are trained by the Science Education Initiative staff, and then bring back their techniques to the department, where they were based, working individually with multiple faculty to transform their courses and teaching methods. At a large institution like UBC or Colorado it was necessary to have 1–2 postdocs embedded in each department, over a period of about 5 years to achieve significant effects. Many of these postdocs were later be hired as faculty in a Teaching Track to help continue the reform for a longer-term impact. This model of imbedded educational specialists is similar to IT support on campus in structure, and is a valuable alternative to a Center for Teaching and Learning, since it can develop faculty teaching expertise from within their own departments and disciplines.

The technique imbedding of postdocs as Educational Specialists helps accelerate the adoption of innovations in STEM departments, as it builds upon "culturally based practices of the organization" and is aware of "core members' self identities" (Wieman 2017). This approach parallels a theory of "Diffusion of Innovation" that describes the ways in which an idea gains momentum through a population or social system (Rogers 2003).

The Science Education Initiative has had dramatic impacts at both institutions, with an analysis of department reports, observations of teaching and TPI data all confirming that 180 and 102 faculty had changed teaching methods at UBC and Colorado, respectively, representing up to 90% in some departments. The SEI provided funding for numerous departments at both institutions at a level of about $1–2 million, and resulted in long-term improvements in institutional culture and pedagogy, with over 71 courses at Colorado and 164 courses changed as of August 2015 (Wieman 2017).

The ultimate goal of the science education initiative however, is to improve science education, and to remedy the current state of affairs where traditional teaching fails to serve students adequately. As Carl puts it,

> Most students are learning that "science" is a set of facts and procedures that are unrelated to the workings of the world and are simply to be memorized without understanding, and they learn to "solve" science problems by memorizing recipes that are of little use other than passing classroom exams. Furthermore, they are leaving their courses seeing science as less interesting and relevant than they did when they started. (Wieman 2017)

The Science Education Initiative is one key element in the changing landscape of STEM education and succeeds by working within the disciplinary cultures of departments.

2.6 Gordon Conference on Discoveries and Physics Education

By 2010, a wide range of innovations in physics and astronomy education were in full bloom. One memorable conference that gave dozens of examples was the Gordon Conference at Colby College in Maine in June 2012 on the topic of "Astronomy's Discoveries and Physics Education" organized by Charles Holbrook and Peter Shaffer of MIT. The meeting was a showcase for ways in which research-based instruction can inspire and motivate students. Research talks on cutting edge astrophysics were given by the leading scientists from Harvard, MIT, and Caltech, and the assembled group of educators and researchers discussed ways to integrate these advances into the undergraduate curriculum. Bob Kirshner of Harvard presented the latest findings on Dark Energy, David Charbonneau of Harvard and Sara Seager of MIT shared the latest discoveries of exoplanets, and Tony Tyson of UC Davis described the next-generation advanced telescopes and instruments. Steryl Phinney of Caltech described how LIGO and space missions finally detect gravitational waves (with the first discovery coming from LIGO just 5 years later!), and give exacting measurements of Dark Energy. My role at the conference was to present new research on nucleosynthesis, quasars and gamma-ray bursts and share how these findings were included in my undergraduate astronomy classes.

Some of the physics education groups include the University of Colorado Physics Education Research group (PER), the University of Arizona Astronomy Education

Research group, led by Ed Prather, and the University of Oregon Physics Education project led by Greg Bothun. Dan Riechart of UNC Chapel Hill described his network of remotely operated telescopes in Chile known as RAPTOR. Andrew West of Boston University described how to use astronomy's vast datasets in the undergraduate curriculum. He was able to work with students to detect and trace the history of thousands of sun-like and smaller stars in the galaxy in introductory astronomy classes using the free Sloan Digital Sky Survey (SDSS) data.

The University of Colorado Physics Education Research (PER) group, represented by Noah Finkelstein, helped frame the issues facing STEM education. Noah presented the challenge facing all of higher education: our institutions are "under threat" by being perceived as being in a "bubble" where the parents and students are faced with rising costs, and are rebelling against what appears to be an unresponsive and overpriced higher education sector. In Noah's words, universities and colleges need to "evolve or die," and must provide a more engaging, interactive and effective form of teaching to survive. Noah cited data from the PISA rankings, where the US ranked 21/30 in science and 25/30 in math. Helping reverse the decline of US students in science and math is the "Grand Challenge in US Education," according to Finkelstein, and this Grand Challenge has to be solved with ever-diminishing resources, and increasingly under-prepared students.

The University of Colorado PER group is facing this challenge with a "scholarly approach to teaching" where new technologies, new curricula, systematic reform, and faculty development are combined with a "theory of change" and an assessment regime. The goal is to escape the old paradigm of education which sees the student as an "empty vessel" to fill with content. Instead the students need to be actively engaged to construct knowledge, based on a curriculum that draws on the student's prior experience, and enables "construction of knowledge" to occur through active learning exercises—the basic steps that Carl Wieman advocated. The new scholarship of teaching used by the PER group carefully takes into account the student's prior knowledge, attitudes, and other attributes. One tool very helpful for this assessment is known as CLASS—the Colorado Learning Attitude Science Survey (University of Colorado Physics Education Technology 2004). CLASS measures prior attitudes before the class, enabling instructors to tailor their classes to the students, and to monitor how the class changes students' conceptions about science. The CLASS survey has also been modified for use in biology classes (Semsar et al. 2011), and is designed to help measure progress of students toward "expert" thinking by assessing their perceptions and problem-solving abilities at the beginning and end of classes.

Another example of a more active form of physics and astronomy education was provided by Greg Bothun of the University of Oregon. Greg developed interactive applets and web sites that make the classroom like a research laboratory for students (Bothun 2013). Rather than complaining about students looking at their laptops during class, Greg *requires* these laptops for students to use to answer questions in class. Greg compiles and shares students' in-class research "on the fly." By leveraging this technology, Greg is able to provide a data-centered and inquiry-based class instead of lecture. As students are answering questions based on their own in-class research, conversations and debates in class provide a lively and engaging atmosphere. Greg's class is also taught without a textbook (Greg regards those as "unnecessarily authoritative"), as he prefers to promote "synthesis" within a class based on student-led inquiry.

Greg presented a few examples of software that he has found effective in his physics and astronomy classes. One is a realistic simulation of the actual observations possible with Galileo's telescope, that allows students to answer the question: "why didn't Galileo derive Kepler's laws?" By taking their own simulated telescopic data, students are able to see precisely how the observational uncertainties prevented Galileo from obtaining the needed precision for finding Kepler's laws. Another computer exercise uses actual stellar and galaxy spectra to measure wavelengths and Doppler shifts, and then uses these measurements to attempt to detect an extrasolar planet. The simulations provide realistic amounts of noise and uncertainty to simulate real data. Students then apply statistics and fitting programs to recreate the discovery of an exoplanet. Additional "Dynamical Astronomy" applets from Chris Mihos of CWRU enable students to crash galaxies together using an n-body simulation, and measure dark matter content in galaxies by simulated measurements of galaxy rotation curves (Mihos 2013). Greg is a strong advocate for creative writing assignments within science courses—having students write reflective essays, policy "white papers," press releases about their "exoplanet discoveries" from the simulations, and collaborative video reports. The mix of inquiry, writing, and original projects livens up the class, and engages the students' talents and creativity better than simply working through equations.

2.7 Active Learning in Biology

While the above examples may give the impression that these active and peer learning techniques are exclusive to physics and astronomy, in fact they are being adopted across all STEM fields, and even outside of STEM. Hundreds of biology instructors have incorporated active and problem-based learning in their introductory and other courses, and these initiatives have been studied and found to increase both engagement and performance of students in introductory biology (Armbruster et al. 2017). The use of Scientific Teaching in biology courses has been developed through training at the HHMI Teaching Fellows program, which has trained over 63 fellows, developed over 35 different sets of instructional materials, new online learning modules, and have enabled dozens of courses reaching thousands of students to incorporate more active learning techniques (Miller et al. 2008). More institutions and departments are adopting the combination of new types of biology curriculum, more effective assessment and active pedagogy, which is having a lasting impact on biology education.

To learn more about the evolution of active learning within biology education, I talked with Kathy Takayama, Senior Science Education Fellow at HHMI, and formerly a Biology professor at Brown University and Director of the Columbia University Center for Teaching and Learning (CTL). Currently, Kathy is directing the

HHMI inclusive excellence program which seeks to increase the diversity of students in STEM and transform institutions to have structural change to improve inclusive STEM education (HHMI 2019). Kathy mentioned that in biology many of the sub-disciplines have launched their own efforts to improve teaching, often via professional development workshops for faculty. The American Society for Microbiology (ASM) is one such group that has had a major impact on biology education. One program is the ASM/NSF Biology Scholars program, which includes an online course for discipline-based education research (DBER) to improve undergraduate biology. The online course has trained over 300 educators in biology and also provides many of the authors and reviewers for the journal JMBE which creates a platform for disseminating innovations in biology education (ASM 2019). ASM also sponsors an annual conference for undergraduate educators to build more effective biology education. Kathy was part of the early ASM education initiatives, as well as a faculty development program known as BioQuest.

Over the past 30 years, BioQuest has built a large community of biology educators dedicated to "effective uses of inquiry, technology, and participatory models of curriculum development." BioQuest began in 1986 by John Jungck from Beloit College, who promoted a new educational philosophy known as the "Bioquest 3P's of investigative biology: problem posing, problem solving, and persuasion" (Bioquest 2019a, b). Workshops for faculty taught new methods of integrating technology and more active learning approaches into their biology courses, initially centered around a set of Macintosh BioQuest computer modules in the 1980s. The program evolved from a software-based education initiative to a broader consortium of biology educators who participated in workshops in the 1990s and beyond, and who have produced dozens of influential articles and curricula in a variety of STEM fields (Bioquest 2019b).

2.8 Creating Research-Based Environments in Biology Courses

Jungck's philosophy of the "3P's" arose from a growing awareness of the "unrealistic" approach to science learning that had "metastasized" throughout the courses in the 1980s. The static textbook, dry labs, and boring lectures were giving a misleading and dated approach to the rapidly changing subject of biology. In a 1988 paper, Jungck stated, "Our courses must help students understand how biologists: perceive the world; pose questions; pursue the problems from those questions; and, ultimately, persuade others of the value of their solutions" (Peterson and Jungck 1988). This shift toward the 3P's provided a more "realistic" form of science education which "captures the open-ended essence of science as it is practiced: problems must be both posed and solved by the problem-solver." While the early computer exercises were only able to provide a limited simulation of "realistic" biology, new research-based biology exercises employed the full array of technologies used in modern genomics in both high school and introductory college classrooms. Bioquest identified the essential ingredients of active learning, which includes the requirement that students are asking their own questions, posing problems and finding solutions. The Bioquest program also placed a strong emphasis on the third P of "persuasion" which enabled students to communicate their findings in ways that overcame objections from skeptical peers (Peterson and Jungck 1988).

Malcolm Campbell, who was part of these early ASM biology education initiatives, helped develop the first undergraduate research-based genomics curriculum at Davidson College, known as the Genome Consortium for Active Teaching (GCAT). The GCAT takes an even more "realistic" approach to biology education, and enables students in introductory classes to gain research experience sequencing genes and then submitting their results for publication in a database and presentation at research conferences. Starting in the 1990s with a small set of students at a liberal arts college, the GCAT program grew to include hundreds of faculty at over 130 campuses, reaching over 10,000 students. An HHMI grant allowed for DNA microarrays or "chips" to be available for only $50 to students interested in participating. Combined with custom software, these chips allow for students to conduct their own experiments using advanced genomics techniques. Faculty workshops were also created to develop a large community of GCAT faculty which grew from 20 faculty in 2000 to over 70 faculty by 2006. This type of "active teaching" in GCAT has been validated from the thousands of students who have continued in careers in science and from survey data that documents changes in student attitudes about science. The GCAT program also aligns well with the recommendations of the Bio2010 report (National Research Council 2003), which emphasized faculty development in genomics, blending mathematics and more authentic research-based laboratory experiences into the cell biology curriculum, and increasing the number of students from under-represented minority groups participating in science (Campbell et al. 2007).

The techniques learned from the GCAT initiative helped inspire the effort known as SEA-PHAGES, which integrates coursework with research in a two-semester, discovery-based undergraduate genomics research course. In the SEA-PHAGES course, students find and name their own bacteriophages from soil samples. The SEA stands for the Science Education Alliance which helps give students access to research early in their academic careers. The PHAGES refers to Phage Hunters Advancing Genomic and Evolutionary Science (HHMI 2019). During 2017–18 the program included over 4800 students from 112 different colleges and universities in the SEA Symposium, where students can share and discuss discoveries. Like the GCAT initiative, SEA-PHAGEs responds to a number of high-level recommendations from organizations of science and government urging for a stronger workforce in STEM fields, and for replacing conventional laboratory courses with discovery-based research courses and authentic research experiences (PCAST 2012; AAAS 2011).

SEA-PHAGES enables both high school and college students to contribute to research by isolating novel bacteriophages, and then sequence the genomes and compare them with a large genomics database to look for new discoveries. The research regularly produces valuable research findings from students in the fields of bacterial pathogenesis, global ecology, and other sub-fields. Because the phages have compact genomes, the task of sequencing and annotation is within reach for introductory students. The HHMI sponsored a scaling of the SEA-PHAGES program to reach 4800 students at 73 institutions over a 5-year effort, including new integrated laboratory exercises, and a SEA-PHAGES research symposium. A parallel faculty development program has trained faculty in new genomics techniques and provides opportunities to share research findings and teaching tips through regular conferences and workshops. Since 2008, SEA-PHAGES students have isolated more than 3000 new phages and have contributed to hundreds of research articles. Students in the program reported large gains in learning, motivation and more positive attitudes about STEM. The SEA-PHAGES program has also significantly improved the grades and retention of students in STEM, compared to students in more traditional biology courses (Jordan et al. 2014).

2.9 Microbes to Molecules and Tiny Earth

Another pioneer in the growth of the research-based biology education is Jo Handelsman, the Director of the Wisconsin Institute for Discovery (WID), a meeting place for interdisciplinary science research and new educational initiatives at the University of Wisconsin. The WID includes a Data Science Hub, a Multi-Omics Hub (for exploring the relationship between genes, environments and life experience), and a Illuminating Discovery Hub (which forms partnerships for outreach and communicating science to the public). Jo's current education initiative is known as TinyEarth (WID 2019a, b) and involves a research-based discovery class where students collect and culture new bacteria which are then analyzed for anti-biotic properties. The chemicals from these bacteria which inhibit the growth of other bacteria are extracted and analyzed, and the hope is that these chemicals can form the basis for new medicines. The program is an outgrowth of Jo's revolutionary course entitled "Microbes to Molecules" which she developed as an alternative research-based introduction to Biology at Yale University. Through the WID, Jo has been able to provide a base for scaling up the course, which now reaches over 10,000 students in 41 U.S. states and 14 countries, creating "a network of instructors and students focused on crowdsourcing antibiotic discovery from soil." The Tiny Earth program includes two instructor training workshops each year, as well as a large database for recording and storing student data online, free genome sequencing for isolates submitted by students, and a Chemistry Hub for helping accelerate the development of new antibiotic medicines (WID 2019b). In addition to helping to discover new antibiotics, students actively contribute to research and present their results at an annual TinyEarth Symposium. The low cost for starting up a program with Tiny Earth (about $600 per class), and the expanded array of curriculum materials and support, assure that this effort will continue to grow and reach even more students in the coming years.

The Tiny Earth project is more than a particular STEM curriculum, however, as it builds on Jo's theory of "Scientific Teaching" in which active learning is a central component to enable students to individually develop their own knowledge, and to help diversify science. Jo described this approach to her teaching in a 2012 interview:

> The aim of scientific teaching is to reach a diversity of students. Classical scientific education reaches only those students who think like the professor. Scientific teaching broadens the cognitive approach, so that more creativity is brought to bear on the way science is presented and communicated. Cognitive research shows us that learning is a deeply individual process, and each person has a unique and idiosyncratic way of developing and constructing our own knowledge. Scientific teaching addresses these unique ways of learning with a general class of activities known as active learning. (Handelsman 2012)

Jo believes that introductory science courses need to develop a student's individuality, creativity, and provide early research experience. These type of research-based introductory courses help students "identify as scientists" and bring them into a research community where they are able to receive support and a social component to their work.

As educational research and teaching practice improves, a growing appreciation of the role of active learning in science education has emerged in all of the STEM fields. Increasing numbers of courses have been developed using active learning across the STEM disciplines, which has provided a large dataset to enable more systematic study of the ways in which the active pedagogy promotes learning. One recent "meta-analysis" used data from 225 different educational research projects to quantify the increases in student learning and performance in active learning courses. The study used combinations of measurements that included examination scores, concept inventories and other assessments across thousands of STEM courses. The results indicated that active learning approaches result in statistically significant increases in examination scores across all STEM disciplines, and that found that "students in classes with traditional lecturing were 1.5 times more likely to fail than students in classes with active learning" (Freeman et al. 2014). These data give increasingly strong signals to educators that in order for STEM education to be more effective and inclusive, peer and active learning techniques are essential. As our knowledge of STEM education improves, science educators have greater appreciation of the need for more humanistic approaches toward science education that recognize the diversity of student learning, that provide inclusive and active learning, and that promote student's self-concept as experts and as scientists. This paradigm shift toward a student-centered active learning approach has ushered in a new age of STEM teaching and learning—no less revolutionary than the Copernican Revolution of astronomy centuries earlier.

References

AAAS. (2011). *Vision and change in undergraduate biology education—A call to action*. Retrieved January 14, 2019, from http://visionandchange.org/files/2011/03/Revised-Vision-and-Change-Final-Report.pdf.

Armbruster, P., Patel, M., Johnson, E., & Weiss, M. (2017). Active learning and student-centered pedagogy improve student attitudes and performance in introductory biology. *CBE-Life Sciences Education, 8*(3).

ASM. (2019). *Improving undergraduate biology education based on research in science learning*. Retrieved January 14, 2019, from https://www.asm.org/Courses/Education-Research-Course.

Bioquest. (2019a). *The BioQUEST curriculum consortium*. Retrieved January 14, 2019, from https://bioquest.org/about-us/.

Bioquest. (2019b). *History of BioQUEST*. Retrieved January 14, 2019, from https://bioquest.org/history-of-bioquest/.

Bothun, G. (2013). *The electronic universe*. Retrieved January 14, 2019, from http://zebu.uoregon.edu/.

Campbell, A. M., Ledbetter, M., Hoopes, L., Eckdahl, T., Heyer, L., Rosenwald, A., et al. (2007). Genome consortium for active teaching: Meeting the goals of BIO2010. *CBE-Life Sciences Education, 6*, 109–118.

Crouch, C., & Mazur, E. (2001). Peer instruction ten years of experience and results. *American Journal of Physics, 69*, 970–977.

Freeman, S., Eddy, S. L., McDonough, M., Smith, M. K., Okoroafor, N., Jorda, H., & Wenderoth, M. P. (2014). Active learning increases student performance in science, engineering, and mathematics. *PNAS*, 201319030. https://doi.org/10.1073/pnas.1319030111.

Hake, R. (2002). Lessons from the physics education reform effort. *Conservation Ecology, 5*(2), 28.

Hake, R. (2005). Interactive-engagement versus traditional methods: A six-thousand-student survey of mechanics test data for introductory physics courses. *American Journal of Physics, 66*, 64.

Halloun, I., & Hestenes, D. (1985). The initial knowledge state of college physics students. *American Journal of Physics, 53*, 1043.

Handelsman. (2012, September 18). Handelsman talks scientific education. *Yale Daily News*. Retrieved January 14, 2019, from https://yaledailynews.com/blog/2012/09/18/handelsman-talks-scientific-education/.

Hestenes, D., Wells, M., & Swackhamer, G. (1992). *The Physics Teacher, 30*, 141–158.

HHMI. (2019). *Inclusive excellence*. Retrieved January 17, 2019, from https://www.hhmi.org/developing-scientists/inclusive-excellence.

Jordan, T., et al. (2014). A broadly implementable research course in phage discovery and genomics for first-year undergraduate students. *MBio, 5*(1), e01051–13. Retrieved from https://doi.org/10.1128/mbio.01051-13.

Kuhn, T. S. (1996). *The structure of scientific revolutions*. Chicago, IL: University of Chicago Press.

Lambert, C. (2012, March–April). Twilight of the lecture. *Harvard Magazine*. Retrieved January 14, 2019, from http://harvardmagazine.com/2012/03/twilight-of-the-lecture.

Mazur, E. (1996). *Peer instruction: A user's manual*. New York, N.Y.: Pearson.

Mazur, E. (2009). Farewell lecture? *Science, 323*, 50.

Mazur, E. (2013). The flipped classroom will redefine the role of educators. *Interview in the EvoLLLution Online Magazine*. Retrieved January 14, 2019, from https://evolllution.com/revenue-streams/distance_online_learning/audio-flipped-classroom-redefine-role-educators-10-years/.

Mazur, E., & Schell, J. (2019). *Turn to your neighbour—Peer instruction network blog*. Retrieved January 14, 2019, from https://blog.peerinstruction.net/.

Mihos, C. (2013). *Dynamical astronomy Javalab*. Retrieved January 14, 2019, from http://burro.astr.cwru.edu/JavaLab/.

Miller, S., Pfund, C., Pribbenow, C. M., & Handelsman, J. (2008). Scientific teaching in practice. *Science, 322*, 1329.

Moore, T. (2019). *Six ideas that shaped physics—A review of the goals*. Retrieved January 14, 2019, from http://www.physics.pomona.edu/sixideas/philosophy.html.

National Research Council. (2003). *BIO2010: Transforming undergraduate education for future research biologists*. Washington, DC: National Academies Press.

PCAST. (2012). *Engage to excel: Producing one million additional college graduates with degrees in science, technology, engineering and mathematics*. Retrieved from https://eric.ed.gov/?id=ED541511.

Peterson, N., & Jungck, J. R. (1988). Problem-posing, problem-solving, and persuasion in biology. *Academic Computing, 2*(6), 14–17. Retrieved January 14, 2019, from https://bioquest.org/note21.html#anchor81555.

Rogers, E. (2003). *Diffusion of innovations* (5th ed.). New York: Free Press.

Semsar, K., Knight, J. K., Birol, G., & Smith, M. K. (2011). The colorado learning attitudes about science survey (CLASS) for use in biology. *CBE Life Science Education, 10*(3), 268.

WID. (2019a). *TinyEarth-About us*. Retrieved January 14, 2019, from https://tinyearth.wisc.edu/about-us/.

WID. (2019b). *Discovering tiny earth*. Retrieved January 14, 2019, from https://tinyearth.wisc.edu/hello-world/.

Wieman, C. (2007a, September). Why not try a scientific approach to science education? *Change, 2007*, 9.

Wieman, C. (2007b, November). The 'Curse of Knowledge,' or Why intuition about teaching often fails (Vol. 16, No. 10). *APS News*. Retrieved January 14, 2019, from https://www.aps.org/publications/apsnews/200711/backpage.cfm.

Wieman, C. (2015). A better way to evaluate undergraduate teaching. *Change: The Magazine of Higher Learning, 47*(1), 6–15.

Wieman, C. (2017). *Improving how universities teach science—Lessons from the science education initiative*. Cambridge: Harvard University Press.

Wieman, C., & Gilbert, S. (2014). The teaching practices inventory: A new tool for characterizing college and university teaching in mathematics and science. *CBE Life Sciences Education, 13*(3). Retrieved January 14, 2019, from https://www.lifescied.org/doi/full/10.1187/cbe.14-02-0023.

Chapter 3
Theories of Teaching and Learning

Abstract Detailed examples of a variety of theories of learning and course design are described to provide an overview of teaching and learning concepts necessary for more effective STEM education. Examples include Jo Handelsman's work on Scientific Teaching, and classic theories of learning such as Piaget's theories of autonomy, Vygotsky's Zone of Proximal Development, Belenky's theories of learning, and other ideas from social constructivism that can be helpful in STEM education. Multiple taxonomies of learning can be used in STEM subjects to help structure course design, and Bloom's Taxonomy, the Feisel Schmitz Taxonomy, and the Miller Taxonomy are described along with types of learning outcomes useful for course design. The ideas of Constructive Alignment and Threshold Concepts can help structure active learning pedagogy learning objectives that can unify and animate courses in STEM disciplines, and a review of these theories along with examples is provided. Schulman's theory of pedagogical content knowledge and its application in STEM courses is described.

Keywords Learning taxonomies · Curriculum design · Constructive alignment · Learning science · Threshold concepts · Scientific teaching

3.1 A Review of Educational Research and SoTL

Our group of faculty were huddled in suspense as we listened to the presentation by Jo Handelsman, who was one of the most celebrated of Yale's science educators, leader of Yale's Scientific Teaching group, and co-chair of the group that wrote the report "Engage to Excel" for the President's Council of Advisors on Science and Technology (PCAST 2012). We were gathered as the group of inaugural faculty for the new Yale-NUS College in Singapore in 2012, as we were discussing science teaching with Professor Handelsman. Rather than deliver a lecture or a dry Powerpoint about educational theory, Jo offered a demonstration.

Jo explained how it helps to present problem a class can't solve to create engagement, and then by working with a class to help them solve the problem, the classroom will reflect the process and "spirit" of science. A video was shown in which very pure water was lowered in temperature to −6 or −8°C, and a special bacterium was

B. E. Penprase, *STEM Education for the 21st Century*,
https://doi.org/10.1007/978-3-030-41633-1_3

added—*pseudonomis synringi*—which triggered a vial of water to freeze instantly (Martin 2011). The video is dramatic, and the rapid freezing was reminiscent of "Ice 9"—a fictional substance that accidentally caused all the water on earth to freeze in Kurt Vonnegut's book *Cat's Cradle* (Vonnegut 1963). Our group was transfixed by the video, which was presented without explanation, and then we were asked to describe what happened, and come up with hypotheses of why the water froze so quickly. The animation and engagement of our group of Yale-NUS professors was tangible and we spent about 30–40 min debating and discussing all the ramifications of this demonstration in cloud seeding, nucleation of ice, and how this phenomenon might be used to regulate the environment. As it turns out, the size of this bacterium and a protein on its cell wall shaped something like the spacing of an ice lattice makes it nearly perfect for nucleating ice. This bacterium is found in nature growing on leaves, and may be transported up into the atmosphere naturally where it may play a key role in regulating climate. The bacterium is also used by humans for snowmaking, seeding clouds, and industrial processes that require quick freezing. Our discussion roved over all of these topics, and was a great demonstration of active learning.

Our experience as "students" was refreshing and stimulating and reminded all of us how excellent teaching can create an exhilarating environment which energizes students and fosters learning. We all have had experiences of this kind—and every teacher wants every class to be as stimulating. And yet the exact mix of content, classroom process and curriculum design can either bring a class to life—or alternatively create a mind-deadening experience. Each professor begins their career with experimentation, guidance and experience and arrives at their own mix of techniques that work. And yet—would it not be more efficient to have a recipe, and a research-validated theory of learning and pedagogy to guide this journey? Unfortunately, STEM teaching often is at a disconnect with education research which has made amazing progress in elucidating how people learn, and how to structure classes, curriculum and assessments to bring about effective and lively learning environments. My task was to help translate some of this literature into working knowledge that could inform our science faculty at Yale-NUS College, through workshops on active learning, course design and other topics. Through collaborations with Yale University, the National University of Singapore, Yale-NUS became an ideal platform for developing an innovative curriculum for undergraduate education and for also experimenting with new types of teaching and learning environments.

Our meeting with Jo Handelsman was one of many experiences that embodied the practices of a master educator. From her discussion, it was apparent that Jo was committed to a form of "inclusive pedagogy" which allows for a diverse population of students to be engaged in courses to help them persist to complete degrees in STEM fields. The need for inclusive pedagogy came through in the PCAST report Jo had just co-authored which made the case for a million new STEM workers by 2020. Despite this urgent need, the current climate of STEM education is extremely inefficient, causing many students to abandon their interests in science either before getting to college, and after enrolling in their first STEM courses in college. Fully 60% of students intending to major in STEM fields abandon these interests in college, and the numbers are even higher for some demographic groups, with 80% of African

American students leaving STEM fields after college! Much of this attrition comes from a "weed out" mentality that persists within some science departments, where instruction is geared toward the students that excel in competitive examinations, driving away many of the most talented students who otherwise could contribute to scientific advances. By bringing the magic of discovery into the classroom along with more opportunities for reflection and interaction by students, students are more engaged and perform better, making teaching more inclusive. To create such learning environments, STEM instructors can employ a "Scientific Teaching" approach—which when guided by theories of learning and systems for organizing objectives and assessing student progress can bring about much more inclusive and effective classes.

3.2 Scientific Teaching

In addition to creating the magic of discovery within a classroom, the best STEM teachers employ larger structures for organizing courses into an integrated system of learning objectives and assessments, with a teaching practice that is responsive to the "data" collected within classes about the learning of students. This practice has been called "Scientific Teaching" and aspires to the same kind of systematic inquiry in a classroom that faculty are familiar with in their research. As expressed in the opening of the book *Scientific Teaching* (Handelsman et al. 2007):

> The goal of scientific teaching is to make teaching more scientific. Embedded in this undertaking is the challenge to all scientists to bring to teaching the critical thinking, rigor, creativity, sprit of experimentation that defines research. Scientific teaching also posits that teaching of science should be faithful to the true nature of science by capturing the process of discovery in the classroom (Handelsman et al. 2007).

Many of the ideas from Scientific Teaching arise from decades of developments of educational theory. Since the 1920s, John Dewey promoted experience and context as necessary for students to learn new things. From his work came the motivation for a proliferation of laboratory experiences that were attached to lecture classes in science, which formed the basic structure of undergraduate science classes in the twentieth century. Central to Dewey's philosophy was the notion that "education must be conceived as a continuing reconstruction of experience." By developing educational experiences as a "process of living" students would be able to develop interest in their "dawning capacities" (Dewey 1897). Dewey's form of scientific teaching included having students combine "(1) observation of surrounding conditions; (2) knowledge of what has happened in similar situations in the past; and (3) judgement, which puts together what is observed and what is recalled to see what they signify" (Dewey 1938). This form of active experimentation in a classroom enables students to refine their interests and enter into a feedback loop which activates the judgement of the student and directs them toward new purposes. A mirroring of this feedback loop by instructors enables a more scientific form of teaching that enables both faculty and students to progress more efficiently and which enables teachers to draw on their own and other's experiences to refine their teaching practice toward newer and more effective forms of pedagogy.

The informed or "scientific" teaching perspective requires teachers and faculty to develop a greater awareness of the purposes of their teaching, and by gathering data on progress on specific and assessable learning objectives. Scientific teaching recommends a pair of assessments to document learning gains—the "formative" assessment at the beginning of the class, and a "summative" assessment at the end of the class. The combination of active learning within each individual class, an architecture of assessment and learning outcomes, and a system for monitoring and adjusting teaching approaches based on feedback makes for a powerful and effective learning experience within such classes. Building the architecture for a course of this kind has been given the moniker of "Backward Course Design" (Fig. 3.1). The process can be concisely summarized by an instructor asking and answering three questions, which in the following order (Handelsman et al. 2007):

1. "*How will my students be different?*" (here you choose course learning goals based student outcomes).
2. "*How will I know if they get there?*" (here you choose assessments—which can include tests, portfolios, presentations, and other demonstrations of competency).
3. "*What will students do to get them there?*" (here you design learning activities—which should include case studies, problem based learning, peer learning exercises, concept mapping, analytical challenges before class, computer simulations, games, and hands-on activity).

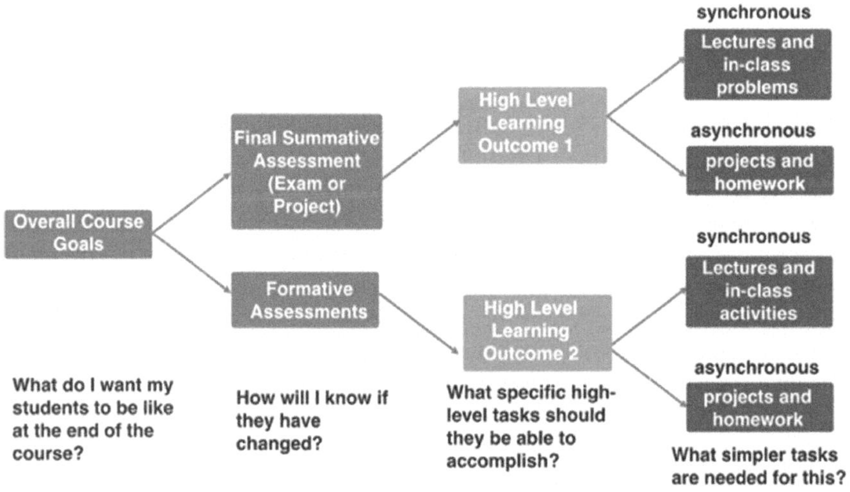

Fig. 3.1 Backward Course Design flow chart showing steps for designing a course with student achievements and learning outcomes as the first consideration, instead of content

Jo Handelsman, Jennifer Frederick, and the Scientific Teaching group at Yale University have been sharing this evidence-based teaching approach nationally since 2004 in Summer Institute workshops sponsored by the National Academies and HHMI. Each year, up to six regional Summer Institutes are offered, reaching over 200 faculty across the US, and these are combined with "mobile" Summer Institutes, in which participants come from the same department or institution. True to the type of teaching shared within the Institute, participants debate case studies, experience active learning settings, and collaborate in groups as they reflect upon and share their teaching with each other. The Scientific Teaching Institute also has its participants develop and share a series of "Teachable Tidbits" that become part of the gathered knowledge of the Summer Institute community. The community has also greatly contributed to the scholarship of teaching and learning as well—as graduates have written over 50 published articles in the educational literature based on the Scientific teaching evidence-based practices (Yale University 2018).

3.3 Theories of Learning

Scientific Teaching places a strong emphasis on "Learning Objectives." More than content-based markers in a course, learning objectives should be a commitment by a faculty to create new capacities within their students to create, analyze, interpret or apply theories to a range of contexts. Choosing these learning objectives effectively requires an awareness of how students learn. Fortunately, research from educational psychologists and philosophers from over a century have greatly clarified the ways in which students combine experience and past conceptions into new understanding. Pioneering work by Piaget and Vygotsky in the early twentieth century recognized the interplay between learning and development in children, which persists in adult learning as well.

Piaget explored the psycho-social roots of learning using a system that described "intellectual norms" that arise within children as they learn and develop expectations based on their experiences. Piaget described these norms or expectations as developing in five stages of increasing generality, beginning with individual *autonomy*, and extending toward causal relationships of wider applicability. These later stages were sequenced as *entailment, intersubjectivity, objectivity and universality* (Palmer 2001). Piaget described how children when they are learning are "constructing" knowledge and developing more extensive and applicable cognitive frameworks throughout life. Awareness of the stages of progression of learning and students' existing cognitive frameworks are vital for learning. In many STEM courses a "diagnostic" test at the beginning of a course or lesson can help probe these "preconceptions" but research has shown that any new information that disagrees with the existing cognitive frameworks that students bring to a class will be resisted, and so systems have to be developed for examining and reconsidering these existing "intellectual norms" of students.

Further descriptions of what is sometimes labelled "social constructivist" learning theory by Vygotsky clarified how social environments can facilitate learning. Vygotsky's best-known contribution is the notion of a Zone of Proximal Development (ZPD) which describes how a learner will only be able to incorporate new knowledge if it is close (or "proximal") to the existing knowledge they are bringing to the class or experience. The process of learning involves moving from interpersonal and social experiences toward an intrapersonal or internalized process—thereby rooting learning in a social context. A simple and operable definition of the ZPD was provided by Vygotsky as "the distance between the (child's) actual developmental level as determined by independent problem solving and the level of potential development as determined through problem solving under adult guidance or in collaboration with more capable peers" (Vygotsky 1978). The role of the teacher is to develop means for recognizing the stage of development of their students, and then providing appropriate guidance to students that is within their ZPD based on this knowledge. This incremental process was later termed "scaffolding," which is a necessary tool needed to guide instructors toward a more responsive form of teaching that enables students to break out from their earlier cognitive frameworks and eventually become self-regulated learners as their learning becomes internalized. It is also interesting to note that the widely adopted approaches of "active learning" discussed in the previous chapter draw much of their philosophical inspiration from Vygotsky's social constructivist theories.

More recent educational research has documented stages of development within students as they progress through the learning process. The work of Belenky et al. (1986) documented how women in her study progressed through five stages or "epistemological positions" as they develop greater agency and confidence in learning. The first of these stages is *silence*—the hallmark of a fearful and disconnected student who needs to be engaged to begin to learn. The second stage is a *received knowledge* position, whereby the authority of the material is unquestioned, and students begin to record and "absorb" information. Unfortunately, traditional STEM teaching through lecturing leaves students in one of the first two stages. Developing confidence and more advanced stages in learners is essential for promoting the internalization of knowledge described by Piaget and Vygotsky, creating what Carl Wieman labels "expert thinking." This more advanced knowledge includes engaging stages of *subjective knowledge*, in which a student's inner voice and feelings are placed in context with the subject, *procedural knowledge*, in which students are given tools to select between competing and sometimes conflicting information. The final stage of learning is the *constructed knowledge* that students internalize, with a full awareness of the ways in which the knowledge is situated in a context that includes their own personal experience.

Students are able to reach this final stage of "constructed knowledge" through appropriately sequenced activities that include sufficient time for reflection and discussion with peers. Belenky's work showed that having the right pace in a learning environment and ample opportunities for students to share their experiences, talk with each other and reflect was crucial for learning—and is the hallmark of inclusive pedagogy. Other elements of an inclusive pedagogy include giving examples of contributors to science that come from a wide range of ethnicities, races and genders, avoiding sporting/military/construction analogies, being aware of a diversity of learning styles (which could be better served by activities such as concept maps and group discussion), being conscious of diversity within discussion groups, and assuring a supportive environment where the voices of all students can be heard (Handelsman et al. 2007).

3.4 Taxonomies of Learning

After knowing more about the "how" of learning, mapping out and connecting the various learning objectives (the "what" of learning) to stages of cognitive development is the domain of the learning taxonomy. Learning taxonomies help classify the types of learning so they can be sequenced in ways that will guide the learning experiences in a course toward more sophisticated and deeper forms of learning. The basic "taxonomy of learning" was developed by Benjamin Bloom in 1956. Bloom's taxonomy describes a ranked set of cognitive activity levels that reflect increasing amounts of sophistication within a student. The most basic form of activity is simply remembering *Knowledge*, or content. The "rote" form of learning and the traditional lecture works heavily on this level. One step above this is *Comprehension*. At this level, students can demonstrate *Comprehension* of facts and how they connect together. The next level of sophistication is *Application* which enables students to apply their knowledge to actual situations, such as in a laboratory section. *Analysis* tasks enable students to de-construct a situation into smaller parts that connect to general principles and knowledge gained earlier. The top two levels of learning, according to Bloom, are *Synthesis*—where students can compile facts and come up with entirely new solutions or concepts, and *Evaluation*—where students can make reason-based decisions about a set of facts on their own, drawing from previously learned knowledge and outside information (Bloom 1956). This highest level of cognition or learning is often labeled "critical thinking" and is one of the main goals of most liberal arts colleges and universities. Science courses (and all education for that matter!) should aim to reach these higher levels of cognition, by creative assignments, independent projects, and hands-on experiments. Figure 3.2 visually represents the Bloom's Taxonomy, with a revised set of words that connect more easily to learning objectives, which should guide a course design.

Like theories of learning, learning taxonomies have also been refined as new research has sharpened how learning works, particularly in STEM subjects. During a 2013 visit to MIT's Learning Laboratory, I met with its Director, Janet Rankin, who conducts educational research and helps MIT faculty develop cutting-edge courses with evidence-based pedagogy. The MIT group recommended the use of the Feisel-Schmitz Technical Taxonomy of Intended Learning Outcomes (Fig. 3.3) (TLL 2005). In this STEM-optimized learning taxonomy, a student moves from an understanding of basic definitions of concepts toward computations using those quantities, then

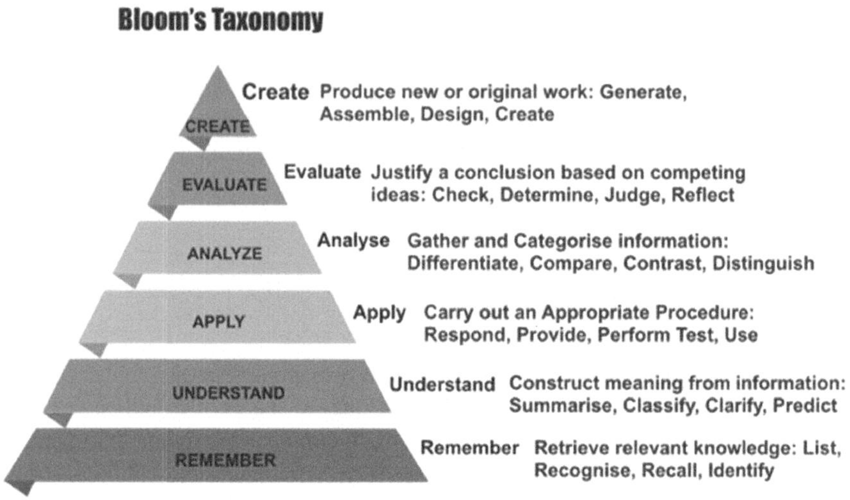

Fig. 3.2 Modified form of Bloom's taxonomy of learning represented in a pyramid diagram, with the highest level of "create" at the top of the pyramid

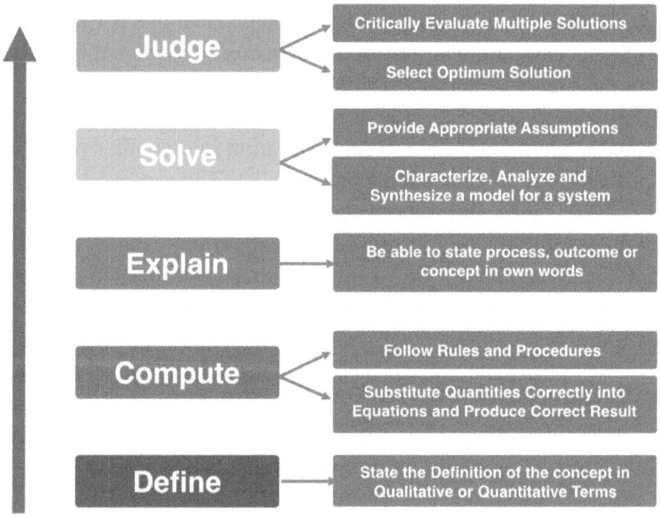

Fig. 3.3 The Feisel Schmitz Technical Taxonomy, an alternative to Bloom's taxonomy for science and engineering courses, showing the increasing levels of sophistication in science and engineering students

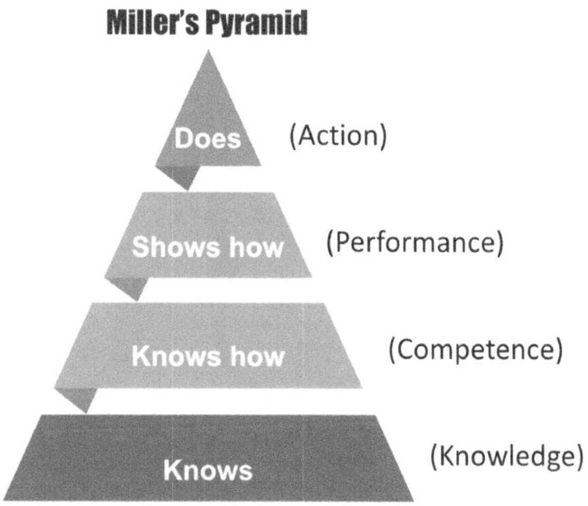

Fig. 3.4 The Miller Pyramid (Miller 1990), a learning taxonomy for clinicians mastering the practice of medicine

toward being able to explain these concepts. The highest two levels of the Feisel-Schmitz taxonomy are to solve a problem by applying these skills—"characterize, analyze and synthesize to model a system" and then to judge— "critically evaluate multiple solutions and select an optimum solution."

Another variation on the Learning Taxonomy has come from the field of medical education, in which learning is classified on the simple basis of how closely the student approaches an actual practicing doctor. This revised taxonomy or "framework for clinical assessment" is sometimes called the Miller Pyramid (Fig. 3.4) (Miller 1990), and ranks learning tasks starting from factual knowledge to demonstrated practice. The ranking in order of complexity is listed below:

- Knows (Knowledge)
- Knows how (Competence)
- Shows how (Performance)
- Does (Action).

3.5 Constructive Alignment

In addition to the development of learning theories and taxonomies, systematic techniques for facilitating learning environments have been developed to match classroom practice to optimize learning. These specific techniques are part of a toolkit that faculty can employ in STEM education to assure many of the elements that have been demonstrated to catalyze learning—active learning, feedback from formative assessments, and opportunities for reflection and application of expertise in collaborative settings.

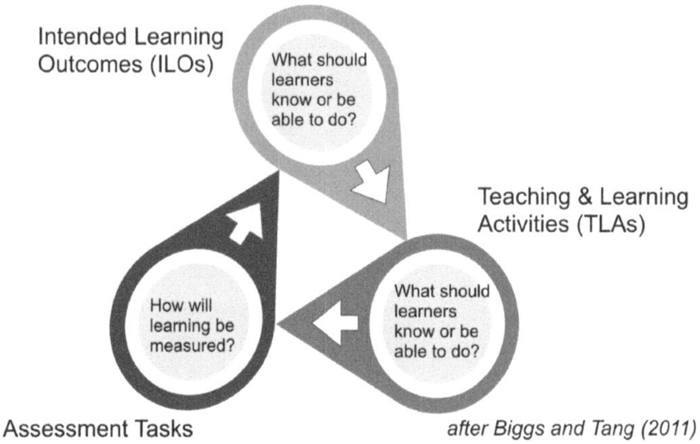

Fig. 3.5 The constructive alignment approach to course design, as proposed by Biggs and Tang (2011)

One very useful classroom practice for organizing the classroom with active learning in ways consistent with the learning objectives of a course is known as "constructive alignment" (Fig. 3.5). Constructive Alignment is based on the idea that students "construct knowledge" and that our courses need to assist this process as a primary goal. This technique is described by one of its main proponents, John Biggs, in summary form:

> Constructive alignment' starts with the notion that the learner constructs his or her own learning through relevant learning activities. The teacher's job is to create a learning environment that supports the learning activities appropriate to achieving the desired learning outcomes. The key is that all components in the teaching system—the curriculum and its intended outcomes, the teaching methods used, the assessment tasks—are aligned to each other. All are tuned to learning activities addressed in the desired learning outcomes. The learner finds it difficult to escape without learning appropriately. (Biggs and Tang 2009).

The constructive alignment approach emphasizes a "systems-based" approach to learning that includes integration of teaching and learning activities (TLA) in the classroom with assessment tasks that respond to the learning objectives, which in the system are labelled as intended learning outcomes (ILOs). The goal is to construct a feedback cycle which includes repeated cycles of meaningful experiences in the classroom, formative feedback and comparison or "alignment" with the ILOs. The three elements are all aligned to reinforce learning so that a student "finds it difficult to escape without learning what he or she is intended to learn" (Biggs and Tang 2009).

The "Integrated Learning Outcome" (or ILO) a central part of Constructive Alignment. These ILO's are designed to produce "functioning knowledge" instead of the more rote-based "declarative knowledge." Functioning knowledge is developed when students "reflect, hypothesize, solve unseen complex problems, or generate new alternatives." Achieving the functioning knowledge requires instructors to choose appropriate Teaching and Learning Activities (TLA's) that otherwise might be described as "active learning." These TLA's require students to "apply, invent, generate new ideas, diagnose and solve problems" (Biggs and Tang 2009). TLA's are combined with assessment, to lead students toward iterative refinement that converge on the desired outcomes. Assessment (from the student's view) is the first, instead of the last, step of this iterative process of improvement in functioning knowledge.

Constructive alignment also works with its own taxonomy of learning known as the SOLO taxonomy ("structure of observed learning outcomes"). Students are led through higher SOLO levels toward more advanced "functioning knowledge" that should be the goals of the Integrated Learning Outcomes. Below is a table categorizing the activities associated with each of the SOLO levels. The left column provides elements of declarative knowledge found on traditional exams (memorize, identify, recite, compare and contrast). The right column includes terms that describe "active learning" SOLO activities that build functioning knowledge. Within class, students can achieve functioning knowledge through sorting tasks, developing illustrations, performing computations, and constructing models and solving problems. The SOLO levels provide a third taxonomy of learning that progress from specific or *uninstructural* tasks toward more complex or *multistructural* and *abstract* tasks (Biggs and Tang 2011).

3.6 Liminality and Threshold Concepts

While many of the tasks of STEM education are able to fit neatly into predetermined learning objectives that can be measured, some of the most meaningful educational experiences in a classroom arise from open-ended dialog and serendipity. Being true to the "spirit of science" requires the classroom experience to admit the unexpected—and allow for spontaneous discovery of ideas that are not part of the lesson plan or learning objective. Most STEM practitioners have limited experience with open-ended discussion, which is the staple of classroom activity in the Humanities and other fields. Facilitating an open-ended dialog in a STEM class can be helped by a framework known as the "Johari window" that categorizes knowledge by the degree to which it is known by the self and others (Fig. 3.6). Ultimately the goal of a discussion is discovery—which involves moving back and forth between the quadrants labelled "discovery through discussion" in which the students and instructors are exchanging unknown information to each other. As with any scientific investigation, there is also the possibility of completely unexpected discovery, and making room for the quadrant of "unknown unknowns" that arise from within the group makes for exciting teaching in which both students and instructors are gaining new insights.

A classroom which is able to foster discussion and openness to new ideas can be considered a "liminal" space. Kathy Takayama, Senior Science Education Fellow at HHMI, and former biology professor at Brown and Columbia, described her approach to liminality in a presentation at the National University of Singapore. She cited Barnett (2004), who noted that "knowing is the position of realizing and producing epistemological gaps" and that educators who hope to teach for an "uncertain

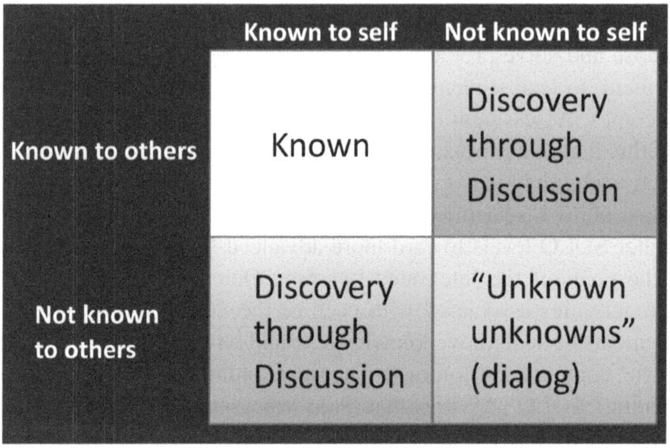

Fig. 3.6 The Johari Window—a tool for identifying quadrants of understanding within discussions, helping to guide student toward "unknown unknowns" (Luft and Ingham 1955)

future" should adopt a schema that includes practices that induce "disciplinary wonder" and "open ontologies for an unknown world." Takayama advocates for a "pedagogy of inquiry" that creates liminal spaces in which students are able to undergo transformation, and be comfortable with risk-taking and exploration. Among the ways to create such spaces include "iterative reflection and critical analysis" as well as new types of student assignments that can validate student uncertainty. Within a STEM course, she advocates using e-portfolios and margin notes within lab books to allow for expressive articulation of the student process of learning (Reichard and Takayama 2012; Takayama 2016).

The very notion of "liminality" is well matched with the educational theory of "threshold concepts" which identify concepts within disciplines that require a drastic reworking one's conceptual framework before crossing the "threshold" of understanding. Threshold concept teaching necessarily creates a liminal classroom since the word "liminal" comes from the Latin root *limen* or "threshold." The threshold concept often can be described as "troublesome knowledge" which requires a new way of looking at the world, but once understood induces a transformative form of learning. Meyer et al. (2006) described both the distinctive features of a threshold concept, and how to use them to avoid overloading courses with less salient content. In summary, a threshold concept is expected to:

(1) Produce transformative learning—as they require an "ontological shift."
(2) Produce irreversible learning—since the understanding has been internalized.
(3) Create awareness of interrelatedness of many phenomena—giving an integrative understanding.
(4) May highlight key locations where a discipline has had a paradigm shift and which continues to inform research in the discipline.

(5) Involves "troublesome knowledge" that might be counter-intuitive or paradoxical.

Teaching with threshold concepts can be thought of as a search for "Jewels in the Curriculum." As Land et al. (2005) described:

> A focus on these jewels, allows for richer and more complex insights into aspects of subjects students are studying; it plays a diagnostic role in alerting tutors to areas of the curriculum where students are likely to encounter troublesome knowledge or experience conceptual difficulty.

Awareness of these "jewels" has become a helpful tool for focusing STEM class time on these most conceptually difficult topics in the curriculum. STEM educators have begun to systematically identify threshold concepts in specific disciplines. Biology educators have key threshold concepts to include the "Testable Hypothesis," along with cellular metabolic processes, cellular size and dimensionality, water movement, genetics, and homeostasis (Moss et al. 2010; Taylor and Meyer 2010). In Physics "Particle Wave Duality" and "Gravity" are key threshold concepts (Meyer et al. 2010; Mead and Gray 2010). Geology and Astronomy both include "Deep Time" and visualizing extremely large numbers as threshold topics (Cheek 2010). Computer Science researchers have identified object-oriented programming and pointers as threshold concepts (Thomas et al. 2010), while Math includes "Limits" among its threshold topics (Meyer and Land 2003). Combining the anticipation of these cognitive obstacles with a classroom that fosters the "liminality" needed to cross the threshold has been a productive advance in STEM education.

3.7 Pedagogical Content Knowledge and Self-similarity

As STEM education begins to incorporate and internalize the results from learning theory and new systems for curriculum design and pedagogical innovation, the disciplines themselves undergo a process of learning that mirrors the developmental phases described above in learners by Piaget and Vygotsky. Professors and STEM departments collectively are encountering "troublesome knowledge" in accepting the need to transform teaching in ways that move away from passive lectures toward pedagogy that inspires and transforms students as they internalize their learning and become "expert thinkers." The theories and techniques described above are generalizable across disciplines as the human process of learning shares certain universalities. The crucial ingredient in STEM or education lies in what Lee Shulman has described as "pedagogical content knowledge" which blends disciplinary expertise with specific pedagogical strategies that allow students to master the cognitive shifts needed for understanding difficult subjects. Schulman described this notion in more detail:

> Within the category of pedagogical content knowledge I include, for the most regularly taught topics in one's subject area, the most useful form of representation of those ideas, the most powerful analogies, illustrations, examples, explanations, and demonstrations—in a word, the ways of representing and formulating the subject that makes it comprehensible to

> others. Pedagogical content knowledge also includes and understanding of what makes the learning of specific topics easy or difficult: the conceptions and preconceptions that students of different ages and backgrounds bring with them to the learning of those most frequently taught topics and lessons. If those preconceptions are misconceptions, which they so often are, teachers need knowledge of the strategies most likely to be fruitful in reorganizing the understanding of learners, because those learners are unlikely to appear before them as blank slates. (Shulman 1996)

In one of Shulman's early works studying how doctors responded to the complex tasks within their profession, he likened the art of teaching to the demanding and rapid thinking practiced by doctors as they rapidly shift between multiple competing hypotheses during a diagnosis and then draw from a vast array of strategies for providing treatment. This analogy was also used in a National Institute of Education report which was entitled *Teaching as Clinical Information Processing* (NIE 1975). In this influential report, the complexity of teaching was acknowledged and included feedback loops between perceptions about students, evaluation of instructional alternatives, navigation within an institutional structure, and integrating method and theory all in the dynamic and changing environment of the classroom. Instead of surgical and pharmaceutical interventions, the STEM educator is drawing upon their disciplinary concepts, pedagogical strategies, and knowledge about student learning drawing upon the theories describe above. In any specific class the instructor would draw from a multiplicity of possible activities to engage students; these might include brainstorming discussions, case studies with students making decisions or taking on roles, in-class peer discussions and short quizzes, concept maps, "one-minute questions," pre and post class questions, and a wide range of experiential and "hands-on" experiments. In the process students are guided toward their own internalized and agile understanding, which in the analogy would correspond toward a state of health and well-being. What "treating" a class, instructors are responding to these multiple feedback loops as they help students comprehend threshold concepts and make progress on a well-selected array of learning outcomes.

The medical analogy for teaching effectively encapsulates both the difficulty and importance of good STEM education, which is capable of deeply affecting students throughout their lives. The analogy falls short in that teachers and students share a more reciprocal relationship in which both are constantly learning from each other, however. As small groups of students test hypotheses, shift their conceptual frameworks, and articulate their new understanding, instructors are also building new understanding as they test their theories of teaching and learning, and learn from their process of discovery in the classroom and from carefully reflecting on how students are learning. The image of students, teachers, groups of teachers and entire professions and institutions learning and improving in their comprehension is reminiscent of a fractal—in which each of these groups embody the same process of learning and growth but at larger scales in time and impact. As groups of students learn from each other, groups of instructors and groups of institutions are also learning—and embody the same "fractal" patters of learning and growing that are based in human learning theory (Fig. 3.7).

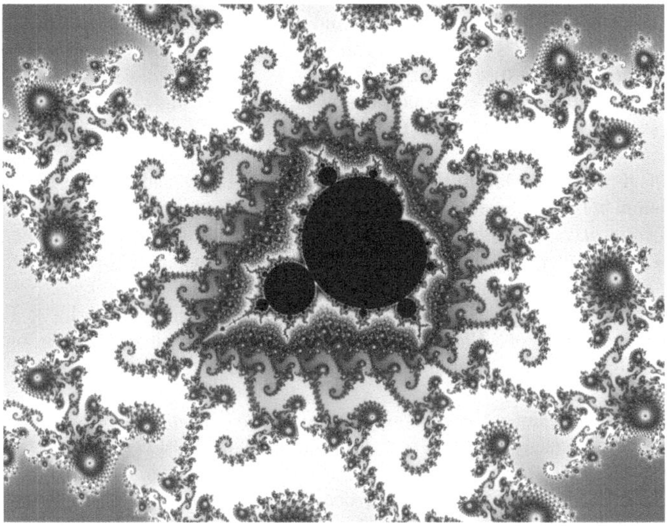

Fig. 3.7 The Mandelbrot set produces beautiful "self-similarity" figures where small structures are identically replicated at larger scales—much as does learning among students, faculty and institutions as they reflect on experience and refine their understanding

References

Barnett, R. (2004). Learning for an unknown future. *Higher Education Research and Development, 23*(3).

Belenky, M. F., Clinchy, B. M., Goldberger, N. R., & Tarulel, J. M. (1986). *Women's ways of knowing*. New York: Basic Books.

Biggs, J., & Tang, C. (2009). *Applying constructive alignment to outcomes-based teaching and learning*. HEA Working Paper.

Biggs, J., & Tang, C. (2011). *Teaching for quality learning at university*. Maidenhead: McGraw-Hill and Open University Press.

Bloom, B. (1956). *Taxonomy of educational objectives, handbook 1: Cognitive domains*. New York: Addison Wesley.

Cheek, K. (2010). Why is geologic time troublesome knowledge? In J. H. F. Meyer, R. Land, & C. Baillie (Eds.), *Threshold concepts and transformational learning* (pp. 117–130). Rotterdam: Sense Publishing.

Dewey, J. (1897). *My pedagogic creed*. New York: E. J. Kellog & Co.

Dewey, J. (1938). *Education and experience*. New York: Simon and Schuster.

Handelsman, J., Miller, S., & Pfund, C. (2007). *Scientific teaching*. New York: W. H. Freeman.

Land, R., Cousin, G., Meyer, J. H. F., & Davies, P. (2005). Threshold concepts and troublesome knowledge (3): Implications for course design and evaluation. In C. Rust (Ed.), *Improving student learning—Equality and diversity*. Oxford: OCSLD.

Luft, J., & Ingham, H. (1955). The JoHari window, a graphic model of interpersonal awareness. In *Proceedings of the Western Training Laboratory in Group Development*. Los Angeles, CA: UCLA.

Martin, M. (2011). *Ice nucleation by bacteria*. Retrieved January 11, 2019, from http://www.youtube.com/watch?v=SenJud3cHLc.

Mead, J., & Gray, S. (2010). Contexts for threshold concepts (I): A conceptual structure for localizing candidates. In J. H. F. Meyer, R. Land, & C. Baillie (Eds.), *Threshold concepts and transformational learning* (pp. 97–116).

Meyer, J., & Land, R. (2003). Threshold concepts and troublesome knowledge 1—Linkages to ways of thinking and practising. In C. Rust (Ed.), *Improving student learning—Ten years on*. OCSLD: Oxford.

Meyer, J. H. F., Land, R., & Davies, P. (2006). Implications of threshold concepts for course design and evaluation. In J. H. F. Meyer & R. Land (Eds.), *Overcoming barriers to student understanding: Threshold concepts and troublesome knowledge*. London and New York: Routledge.

Miller, G. E. (1990). The assessment of clinical skills/competence/performance. *Academic Medicine, 63,* S63–S67.

Moss, P. M., Taylor, C., Hughes, C., Kofod, M., Whitaker, N., Lutze-Mann, L., & Tzioumis, V. (2010). Threshold concepts: Challenging the way we think, teach and learn in biology. In J. H. F. Meyer, R. Land, & C. Baillie (Eds.), *Threshold concepts and transformational learning* (pp. 179–192).

NIE. (1975). Teaching as clinical information processing. In *NIE Conference on Studies in Teaching; Panel 6*. Retrieved January 11, 2019, from https://eric.ed.gov/?id=ED111807.

Palmer, J. A. (2001). *Fifty modern thinkers on education* (p. 38). Oxon: Routledge.

PCAST. (2012). Engage to excel: Producing one million additional college graduates. *Degrees in Science, Technology, Engineering and Mathematics*. Retrieved from https://eric.ed.gov/?id=ED541511.

Reichard, D., & Takayama, K. (2012). Exploring student learning in unfamiliar territory: A humanist and a scientist compare notes. In K. McKinney (Ed.), *The scholarship of teaching and learning in and across the disciplines*. Bloomington: Indiana University Press.

Shulman, L. S. (1996). Paradigms and research programs in the study of teaching: A contemporary perspective. In M. C. Wittrock (Ed.), *Handbook of research on teaching* (3rd ed., pp. 3–36). New York: Macmillan.

Takayama, K. (2016). The value of uncertainty in the learning sciences: Intersections of technology and student engagement. In *Keynote Address to the 2nd Global Learning Council Symposium*, 12–13 April 2016. National University of Singapore.

Taylor, C., & Meyer, H. F. (2010). The testable hypothesis as a threshold concept for biology students. In J. H. F. Meyer, R. Land, & C. Baillie (Eds.), *Threshold concepts and transformational learning* (pp. 179–192). Rotterdam: Sense Publishing.

Teaching and Learning Laboratory. (2005). *Two examples of taxonomies of educational outcomes*. Retrieved January 11, 2019, from https://tll.mit.edu/help/two-examples-taxonomies-educational-objectives.

Thomas, L., et al. (2010). Threshold concepts in computer science: An ongoing empirical investigation. In J. H. F. Meyer, R. Land, & C. Baillie (Eds.), *Threshold concepts and transformational learning* (pp. 241–258). Rotterdam: Sense Publishing.

Vonnegut, K. (1963). *Cat's cradle*. New York: Delacourt Press.

Vygotsky, L. S. (1978). Mind in society. In M. Cole, V. John-Steiner, S. Scribner, & E. Souberman (Eds.), *Mind in society: The development of higher psychological processes*. Cambridge: Harvard University Press.

Yale University. (2018). *Scientific Teaching Summer Institute*. Retrieved January 11, 2019, from https://poorvucenter.yale.edu/teaching/stem-education-and-evaluation/summer-institutes-scientific-teaching.

Chapter 4
Engineering Education Reconsidered

Abstract The new engineering education for the 21st century has been inspired by blending "design thinking" with active pedagogies to provide exciting new learning environments for engineering. Detailed case studies of leading programs of innovative engineering education are provided, including the Center for Engineering at Yale University, California Polytechnic San Luis Obispo's "Learn by Doing" approach, and the innovative Olin College of Engineering with its emphasis on project-based learning. These three programs illustrate how new types of engineering education train students more effectively for the complex and interdisciplinary tasks that today's engineers face in their work. A review of the new Liberal Arts Engineering movement is described, with detailed examples of the leading engineering programs that are blending disciplines from outside of STEM fields into the engineering curriculum to deepen and strengthen engineering education.

Keywords Design thinking · Maker spaces · Project-based learning · Liberal arts engineering · Interdisciplinary curriculum · Engineering education

The new age of engineering education recognizes diversity as a strength within design teams, and within society, and seeks linkages between engineering and liberal arts disciplines to take a more humanistic approach to engineering. The new engineering incorporates "design thinking" which draws from social sciences, humanities and the arts. Three illustrative examples below, from Yale University, Cal Poly San Luis Obispo, and Olin College of Engineering show the range of new approaches being used in engineering education at leading institutions and how they are preparing students more effectively for the complex and interdisciplinary tasks that today's engineers face in their work. These new forms of engineering are being developed—at Yale, a 300 year-old private Ivy League institution, at Cal Poly San Luis Obispo, a public "land grant" university, and at Olin, a 15 year old "start-up" private engineering college. The new age of engineering education is emerging in these institutions and dozens of others, in which many of the old assumptions are being challenged and engineering education is being reconsidered—and improved to be more responsive to the needs of society and to the students.

B. E. Penprase, *STEM Education for the 21st Century*,
https://doi.org/10.1007/978-3-030-41633-1_4

4.1 Yale Engineering and the CEID

An exciting new element in engineering education are hands-on centers where students can design and build their own inventions, with help from skilled staff members using advanced equipment. Stanford opened the *d.school* in 2005, which was intended to foster the type of creativity so abundant in Silicon Valley startups. The d.school included an "operators handbook" that described how "creative ecosystems like the d.school thrive when they are led collectively and guided in an emergent way" (Kembel 2007). The d.school is able to provide a space for "radical collaboration" for people from all disciplines to work on real-world projects. In 2011 Harvard opened an Innovation lab or "Hi" and is also known on campus as the "i-Lab." The i-Lab complex includes 30,000 square feet of high-tech space outfitted like a startup. According to Joseph Lassiter, the faculty chair of the i-Lab, "People can try out their ideas and see if they are worth putting to use" (Harvard Magazine 2011). Yale's entry into this new type of center opened during 2012 and is known as the Yale Center for Engineering Innovation and Design or CEID. All of these centers have emerged in the past decade as a response to a need for a new and different kind of engineering education, one that is more interdisciplinary, more creative and one that reflects "design thinking." The Yale CEID and Yale Engineering program is an instructive case study in the way in which Engineering is being reconsidered.

The Yale CEID is intended to be more than a place for students to work on projects. Its central location, at the base of "Science Hill" close to the undergraduate colleges, provides convenient access that ties together the campus and forms an architectural mid-point between the science and humanities at Yale's campus. Large windows line the exterior walls, allowing pedestrians walking by to look in and see what is happening at the CEID, in its open, light-filled space. The Center's cafe features a huge wall to ceiling LED display that streams interesting patterns and images. The Center's interior is stocked with the shop benches bristling with tools and electronics, advanced 3D printers, laser cutters, a chemistry lab, and a fully CAD-CAM compatible machine shop.

I visited the CEID shortly after its opening in 2012, to learn more about the CEID and Yale's approach to engineering education. Kyle Vanderlick, Dean of the Yale School of Engineering and Applied Sciences (SEAS), described how the CEID is part of the resurgence of Yale's engineering program. Kyle described how Yale's engineering program was the first program to award an engineering PhD in the United States in 1863. This degree was given to Josiah Willard Gibbs, who later went on to become one of the most accomplished American scientists ever—a man so brilliant that Einstein called him "the greatest mind in American history." When the Sheffield School (as it was known) began its engineering program in 1852, it offered a program in civil engineering, then added mining engineering in 1865, and metallurgy and mechanical engineering soon afterwards. Additional programs in electrical engineering and chemical engineering were added in the 1890 and in 1920s, along with industrial engineering in 1921 (Cunningham 1992).

By the 1930s Yale Engineering had grown into six different departments, and in 1932 was constituted as the "Yale School of Engineering." By the 1950s the engineering school seemed disconnected from the larger mission of the university, however, and in the 1960s it lost its accreditation, shrank from graduating 200 to around 10 PhD's per year, and was downgraded from a school into a single department with many faculty leaving. In the following years, Yale engineering struggled, and nearly was disbanded. The resurgence of the program began in 1993 when Richard Levin became Yale's president. To revive the program, Levin appointed nuclear physicist Alan Bromley as Engineering Dean. Bromley was a former White House science advisor, who returned to Yale from 1994 to 2000 to reboot Yale Engineering. From the 1990s onward, Yale's engineering transitioned into an interdisciplinary school that leverages Yale's strengths in medicine and humanities. In a Yale daily news article from 2009, Kyle pointed out the unique aspects of Yale Engineering:

> We have certain strengths because we are small. We are more interdisciplinary, we're more nimble; we have better connections to other parts of campus that other departments of engineering don't have. It's just not that seamless, it's just not that collaborative, at all places. (Letchford 2009)

Yale Engineering today offers interdisciplinary degrees such as biomedical engineering and environmental engineering, and is connected to many other parts of campus, the schools of management, forestry, medicine, and even the humanities. Yale Engineering offers non-traditional courses like "Appropriate Technology in the Developing World." This course enables students to retrofit motorbikes or take surplus gears and sprockets to build devices that can be used by farmers in Africa to generate electricity, pump water, or grind corn and flour. Yale's strong emphasis on "bridging" between humanities and sciences was emphasized by Kyle Vanderlick in 2011:

> Though rarely described this way, engineering is the bridge between the sciences and humanities. Simply put, engineers apply scientific principles to advance the human condition. Their success relies as much upon an understanding of physics and math as an appreciation of history and psychology. (Vanderlick 2011)

The CEID is perhaps the centerpiece of the Yale Engineering school's strategic plan. More than infrastructure and equipment, the CEID is designed to catalyze interdisciplinary work and community among students working in engineering, sciences or the arts. Any Yale undergraduate or graduate student can join the center and work 24 h a day to build, to prototype, and to explore possibilities in a welcoming and fun space. The importance of the CEID to Yale's community was evident in its opening ceremony during February of 2013. The opening featured a wide range of Yale faculty, alumni, industrial leaders (including the Boeing CEO, W. James McNerney, Jr.—Yale '71), and Yale administrators such as Peter Salovey, then Yale's new President-elect. In the opening speeches, the leaders stressed the ways in which the center is hoped to bring students in art, physics, engineering, and even psychology together, since engineering and technology requires this sort of interdisciplinary approach to design products that people enjoy using, and that are economically, ethically, and technologically sound.

After the CEID opening ceremony, I met with Eric Dufresne, the inaugural CEID director. Eric described how the CEID and his personal philosophy of education aligned toward promoting student creativity. Yale has long been a leader in humanities, and has innumerable theatre productions in its Colleges, art studios, and music groups. This same type of creativity is now be expressed at the CEID, according to Eric, and he described how the center can unleash creative energies in science and technology so that students can "realize ideas." Eric believes this type of open-ended, student-driven science education is the future of higher education.

The student-built contraptions were on display at the final presentations of Dufresne's freshman class "Introduction to Engineering Innovation and Design." First-year Yale undergraduates were charged with building devices to solve problems using the resources of the CEID and collaborating Yale faculty. Yale College Dean Mary Miller, and a panel of engineers were also on hand to judge the projects. One group designed a housing for to help biologists research water quality in the Mara River (Tanzania). They presented their device in the form of a one act play, that enabled them to dramatize the challenges that such a device faces from being stomped on by a hippopotamus, and being clogged by river debris. A second team demonstrated how they designed and built a small and portable blood assay system where injection of a small sample of blood into a tiny capillary channel will enable it to respond to light. Their device made use of both the laser cutting and 3D printing capabilities of the CEID. A third team developed a new sensor and software system to enable the huge LED system in the CEID cafe to register the number of people in the cafe in the form of an entertaining display of bouncing balls. The LED system, according to Yale Deputy Dean of Engineering, Vince Wilczynski, is "the largest low-resolution graphic system" and includes a bank of LEDs in a large display that spans two walls and the ceiling of the cafe. The students designed their system using n-body physics code to realistically render the bouncing balls, infrared sensors, and a sound system that registered when people entered or left the cafe. The excitement and fun of this contest and the engagement of these Yale first-year students provides good evidence both of the CEID achieving its goals and of Yale Engineering reaching out to the Yale undergraduates and the larger Yale Community.

The Yale Engineering program and CEID has enabled Yale to make engineering more interdisciplinary, more engaged with the mission of the college, and more relevant to students and to the world. Yale Engineering seeks to produce engineers that have "breadth," "depth," and "purpose." As stated in one of their reports, Yale Engineering graduates students who have broad awareness of the context of their work, with "fundamental knowledge of science and engineering, but also a strong understanding of the complex social, political, economic, and environmental implications that must be considered when developing complete solutions to global problems" (Yale SEAS 2019).

Vincent Wilczynski, the Deputy Dean of Yale Engineering, described how the convergence of these three factors creates a "Y shaped engineer" that combines all three—breadth that includes liberal arts and other subjects, depth in one discipline, and a purpose to applications that solve problems. Since founding the CEID in 2013,

Yale Engineering and Applied Science has continued to grow and offers a summer SEAS Summer Design Fellows program where students are able to invent new devices in 8 weeks and present them to professors. The Biomedical Engineering program has expanded and now includes a number of innovative projects that are collaborations between engineering, biology, mathematics and medicine, and are developing new advanced brain imaging techniques, tissue mechanics and research in systems biology. Additional projects are developing new forms of water treatment using nanotechnology, and computer applications such as cryptography, natural language processing and other artificial intelligence applications. Within the Yale Engineering program is also a strong component of "Green Engineering" which has been integrated into the curriculum, and which informs the design philosophy of many of the students and faculty (Yale SEAS 2018).

4.2 Cal Poly San Luis Obispo and "Learn by Doing"

Another institution that exemplifies many of the emerging best practices in engineering education is California Polytechnic University, San Luis Obispo, or "Cal Poly." The Cal Poly philosophy of "Learn by Doing" has guided their curriculum and institution for decades, and has resulted in some of the most impressive graduates in science and engineering in the country. Cal Poly's undergraduate engineering programs are ranked alongside those of Stanford and MIT, despite their much lower budgets and tuition rates (in-state annual tuition at Cal Poly for 2018/2019 is only $9789!). The US News rankings placed Cal Poly as #1 in Top Public Schools, and Most Innovative Schools, with top three rankings for undergraduate engineering in Computer Engineering, Civil Engineering, Aerospace Engineering, Industrial Engineering and Mechanical Engineering. The Cal Poly physics program was chosen by the American Physical Society as one of the best 21 "thriving" physics programs in the country to be studied as part of the "Spin-UP" report, which praised them for being "friendly and open with a broad, hands-on and can-do approach to physics." Understandably, students in California and beyond are clamoring to get into Cal Poly, and they received 45,000 applications for only 4000 spots in the entering class, giving them a selectivity more like that of an Ivy League University than a Cal State campus!

To learn more about the Cal Poly approach to STEM education, I visited the Cal Poly San Luis Obispo President, Jeff Armstrong, in 2013, who helped set up additional meetings with some of the Cal Poly leaders, such as Enz Finken, Provost, Rakesh Goel, Associate Dean of Engineering, Phil Bailey, Dean of Science, and John Keller, the co-director of CESaME—an education center for science and engineering. My first meeting was with Jeff Armstrong, the Cal Poly President. Jeff comes from a background is in large food animals and their nutrition and reproduction, and is a leading researcher and award-winning teacher in this subject. His practical background is consistent with the engineering and agricultural strengths of Cal Poly. Before coming to Cal Poly in 2011, Jeff worked at Michigan State, Purdue and North Carolina, and was raised on a farm in Western Kentucky. He met with me in his office and was enthusiastic about Cal Poly since in his words, its "Learn by Doing" philosophy has led to a "renaissance of the practical." Jeff has three main goals—to improve graduation rates, to improve the campus climate for diversity and inclusivity, and to increase fundraising and partnerships with private industry.

Cal Poly has a much less diverse student body than the rest of the Cal State Universities, with less than half of the students coming from traditionally underrepresented minorities (Clark 2018). To improve this situation, several new programs were put in place. The Cal Poly EPIC summer program enables junior high and high school kids to explore "Engineering Possibilities in College." Minority and first-generation students were also recruited with on-campus weekends (both an "Open House" and "Polycultural Weekend"). This program has expanded to include four annual sessions accommodating over 640 participants each year by 2018 (Cal Poly EPIC 2019). A Summer Institute provides a solid orientation to the campus culture to pre-frosh students, and many of these students are eligible for a wide range of scholarships. The Summer Institute allows students to take six units before starting, giving them a head start on their academics, as well as to build study skills and take part in a series of social activities and advising opportunities (Cal Poly EOP 2019).

As a first-generation college student himself, Jeff has a solid understanding of the importance of improving the experience of minority and first-generation students. In all of these efforts, Jeff is guided by his four guiding principles—student success, learn by doing, reaching for excellence through continuous improvement in all things, and to offer a comprehensive polytechnic university which fills a niche and offers great value to the Cal Poly students. The clearly defined educational philosophy of Cal Poly, and the methodical priorities and guiding principles provide a solid and pragmatic basis for academic excellence—consistent with the overall philosophy of the institution.

One key figure who helped build Cal Poly's success over the years is Phil Bailey, who until recently was the Dean of Science and Mathematics at Cal Poly. During his 44 years on campus, Phil tied together all the rich academic and programmatic offerings with an emphasis the human side of studying engineering and science. Before his retirement in 2017, Phil was a father figure to many of the students and communicated regularly with the entire entering class with a series of letters to "give ideas about how to navigate the first quarter of college." Phil hosted many of the students in his home, and his warm approach has helped many generations of students feel comfortable and welcome on campus. Phil placed student success as one of the best measures of teaching ability and helped faculty redesign courses to improve the performance of students in intro courses. One example he cited was a re-organization of Math courses, which led to a decrease in the D/F rate in introductory math from 70 to 20%. Phil described how many of the Cal Poly professors are moving toward new types of pedagogy, and that more than 20% of the Cal Poly curriculum is in "non-lecture" format. The Cal Poly "Learn by Doing" approach places a heavy emphasis on lab work, and Phil helped build a new studio chemistry curriculum, where lecture and lab are integrated in all of the chemistry courses. Even though a lab can be more expensive than a lecture course, the gains in learning have convinced Cal Poly to expand the laboratory program, even in the face of decreasing state budgets.

Phil believes that students need to be persuaded and taught how to study and cited an NSSE study that showed that 40% of the students in college study less than 10 h per week. Phil's mantra is "Study 25–35 h per week!" Phil started a "25–35 program" to promote this idea and emblazoned the slogan onto refrigerator magnets he cheerfully handed out to students. Many students were actually not aware of the necessity of longer study hours for success, as their high schools often did not demand the same level of preparation and work as Cal Poly does from its students. Phil's work increasing student success emphasized not only studying, but compassion and concern for the students. His approach had two key elements. (1) Treat each student as an individual, and (2) Seek out students at risk and find out what is wrong. This humanistic approach to engineering education has provided countless students much-needed support as they entered into their studies.

Since Cal Poly is a practical place, it not only educates students for jobs—it also helps students create new companies and products. A new Center for Innovation and Entrepreneurship gives students chances to compete for prizes, meet mentors, and work in teams through their "Hothouse." During 2012, Cal Poly supported seven student-designed companies in this summer program, including teams that will develop new vegetable snacks, implement software for splitting bills among apartment roommates, design vertical gardens for buildings, and make a new form of lightweight concrete using cellulose and cement. By 2018 the program has expanded to reach hundreds of students through programs such as their "Hatchery," "Innovation Quest" and "Startup Weekend Program." The program now includes both an incubator and accelerator, which allows for businesses to be established and to grow, leveraging the skills of students and faculty at Cal Poly. Among their "alumni" are companies like "Flume" which conserves water by innovative new fixtures to gauge and control water usage. By 2018, 89 companies have started at the CIE, along with 61 companies in the "Hatchery," and these companies have raised over $136 million in startup funding (Cal Poly CIE 2019).

Two other notable programs at Cal Poly embody the hands-on and high impact approach of this institution include the CeSaME center, and the CubeSat program. Cal Poly's CeSaME center is an acronym for the "Center for Excellence in Science and Mathematics Education." The center runs a number of exciting workshops for teachers, including the STEM Teacher and Research program, which matches high school science teachers with researchers at national laboratories funded by NOAA, DOE, and NASA for a summer project. In 2013, the STAR program provided 290 research opportunities for high school teachers within 15 different research sites. CeSaME also houses the "Learn by Doing Lab" which is a facility that gives Cal Poly students interested in teaching a space to try innovative new hands-on activities with visiting school classes (Cal Poly CESAME 2019). Teachers can also check out the facility for use in their classes. The CESaME center also participates in a number of initiatives, including math tutoring for area K-12 students, administering the NSF-funded Noyce scholarship program for students interested in teaching science or math, and participating in the Central Coast STEM Collaborative, and the California STEM learning network. By 2018 the CESaME programs expanded to include multiple cross-campus initiatives as well as state-wide teacher education and professional development programs. These include the STEM Teacher and Researcher (STAR) summer research internship programs for science and math teachers, and the Math Science Teacher Initiative grant. Cal Poly's emphasis on assuring student success in STEM fields, and increasing access to these fields from diverse populations of students is well-served by these initiatives, which also have large impacts beyond the Cal Poly campus.

The Cubesat program at Cal Poly exemplifies the type of cutting-edge undergraduate work that is possible in an institution dedicated to "Learn by Doing" (Fig. 4.1). By 2013, over 100 Cal Poly students have helped develop and launch over 32 "CubeSats" which are tiny cubical satellites (the face is about the size of a CD jewel case $-10\ \text{cm}^2$) packed with electronics for studying space.

NASA wrote a $5-million contract with Stanford University and Cal Poly to extend the CubeSat program to more universities in 2013. CubeSat satellites were launched using an innovation known as the P-POD device that integrates these cubes to existing launch vehicles. This device enables students to design and build self-contained space instruments that can piggyback on other NASA launches. The maximum mass of these satellites (also known as "picosats") is 1 kg, making a design challenge for students to devise instruments small and light enough to fit in the compact package. Cal Poly professor Jordi Puig-Suari, the founder of the program, described his goals

Fig. 4.1 The Cal Poly "CubeSat" being tested in the environment near campus

for the program: "It used to be space was a very small club because it was so expensive. Now, all these young students play with it in school, build a cubesat, make a bunch of mistakes and then say, 'I could do XYZ with a cubesat'" (Werner 2018).

Since 2013, the CubeSat has helped spark a "New Space Revolution" and can be developed in a number of configurations, which are specified by the number of cubes linked together—with a 3u Cubesat consisting of 3 of the cubes forming a longer and more sophisticated system. Cubesats in 1u–3u and even 6u are now being designed by students across the country, and a Cubesat 101 guide and annual workshop is offered by Cal Poly to any interested developers. A variety of "new space" companies are now available enabling launches from private companies such as NanoRacks, Tyvak and Spaceflight Industries, who contract for launch using private companies such as SpaceX and Orbital, who have developed their own rockets and launch capabilities. Cal Poly on its own has launched an entire fleet of satellites, with an impressive listing of the missions on their website. Starting with the CP1 satellite in 2006, and continuing on through 2018 over 26 missions listed on the website (Cubesat.org 2019), Cal Poly's space program is more advanced than many countries!

4.3 Olin College of Engineering

Olin College was founded in 1997 and opened its doors in 2002 in a comfortable suburb of Boston, where it provides an example of a dazzling successful start-up institution built around the idea of the centrality of experiential and active learning in engineering (Fig. 4.2). A case study in IEEE Spectrum magazine described the "Olin Experiment" with the subtitle "Can a tiny college reinvent engineering education?" and gave examples of Olin's approach (Guizzo 2006). In the case study, students were charged with building a device to crawl up a wall, and faculty watched, and then provided help and content as needed in the design and build process. Professor Benjamin Linder, Associate Professor of Design and Mechanical Engineering, is described in the case study:

> "Linder, who studied product design at MIT before becoming a mechanical engineering professor at Olin, tells me that his course is a 'bio-inspired introduction to design.' The class includes two projects during the semester. The first is a mechanical hopper. Students consider click beetles, springtails, spittlebugs, and fleas. They study how the insects propel themselves, and they use that knowledge to design their own hoppers, he explains. 'Did I show you the damage one did to the ceiling?' The second project, more challenging, is the glass wall climber, which the students make out of plastic pieces, electric motors, pneumatic actuators, and suction cups." (Guizzo 2006)

The idea of having students embark on a design project first, and then learn relevant content while building something seemed very exciting, but also daunting and somewhat dangerous. This risk-taking approach underlies the institutional culture of Olin, and they try to maintain a fresh and experiential approach to all of their classes. During a site visit in May 2013, I met with Olin's Founding President, Richard Miller,

Fig. 4.2 Olin College campus—which includes a circular courtyard surrounded by the academic buildings and student center (right). Students occupy a table for the "Build Day" festivities which were happening during a site visit in 2013

two of its founders, Charlie Nolan (then Dean of Admissions) and Steve Hannabury (CFO), along with the Dean, Vince Manno, who joined Olin in 2011.

Olin's President, like the campus, appears at first to be conventional, but under the quiet exterior is a revolutionary. Richard Miller spent the better part of an hour telling us about the founding of Olin, and the excitement and the challenges of the early days. Richard was "employee #1" of Olin, and was hired soon after writing a white paper on engineering education while he was Dean of Engineering at the University of Iowa. Lawrence Milas, President of the Olin Foundation, called Richard and asked him what he would like to do differently in engineering education. Richard then visited Milas to discuss his ideas in person, and wrote up an extended letter about what he thought Olin should do. This letter, which Richard had put into a drawer after his visit, later became the blueprint for the new Olin College.

The Olin Foundation called Richard back, and soon he was fully engaged hiring new faculty to help found the new Olin College. Richard first sought the advice of many senior faculty who were frustrated with the state of engineering education and then hired many younger faculty with passion to make the new college a reality. Within a year, Richard had hired about a dozen of the first employees, including four of the Vice Presidents. Richard viewed diversity in thought to be absolutely critical in the ranks of faculty. With the diversity came challenges, as many of the faculty had very strong personalities and were passionate about their ideas. Richard credits "having a long fuse" as being very helpful in these early days. He also likens founding a college to being something like a "country doctor" as you have to do everything.

In the first year, 2001, a group of 30 students (selected from over 600 applicants) helped the 15–20 faculty test out some of the concepts in what was called "The Partner Year." During that year, the small group of pioneering faculty and students tested out different class topics and teaching approaches together. The first official batch of 75 students arrived in 2002 and were offered free tuition. The students and faculty were both excited, but the ambitious curriculum was designed around short half-semester courses known as "quamesters." The pace at first was too quick—and caused students to lose sleep building things until 1–2 AM. Within a few weeks some students were exhausted and faculty too were frustrated with the rapid pace of the first short courses. After the winter break, the curriculum was re-structured to include three linked courses—which would allow for instructors to work together to be sure that the engineering projects were integrated with basic concepts of the academic subjects such as calculus, physics and engineering design (Marcus 2002).

Based on advice from Joe Platt, the Harvey Mudd College President, Olin grew slowly with two "big ideas." The first principle was making sure that the students are excellent—"10 excellent students are better than 40 average students"— even if this meant lower enrolments in the first years. The initial students were offered a full scholarship to attract the best and brightest, who might otherwise go to MIT or Stanford. Olin's first Provost was a former Harvey Mudd mathematician named Mike Moody, who helped recruit and hire some of the first faculty. The excellent students in turn attracted some of the best faculty so that Olin College now receives 150–200 resumes for each open position, even though these faculty jobs come without the assurance of tenure. The second principle was to be mindful of the tendency for becoming conservative and to continually strive to keep the curriculum, the instruction and the institutional culture fresh. New generations of faculty and new generations of students all want to share the sense of being at a new institution, and the leadership strives to meet this expectation.

Olin's curriculum has succeeded in gaining national attention, with articles such as the 2007 New York Times Magazine article "Re-Engineering Engineering," and was hailed by the Kaplan review as one of the "new Ivies" (Schwartz 2007). Recent rankings for Olin College include #3 best undergraduate engineering (from US News), inclusion in Princeton Review rankings with #3 for "Professors Get High marks" and #6 for "Students Study the Most." In 2013, Olin also received the prestigious Gordon Prize from the National Academy of Engineering to recognize innovation in engineering education.

The Olin College curriculum stresses that the design process is based in people, and that this "realization process" requires engineers to understand that design requirements come from the needs of the people using the object, and this informs how it is built, and how it is marketed. Olin uses an approach called "user-oriented collaborative design" which includes an awareness of the social values that play a key role in effective design. The method is rooted in close observation of the values and lifestyles of people using the designs:

> Students observe and engage people to develop a deep understanding of their values and the patterns of their lives, and then develop detailed concepts and models of authentic new products and services. They work collaboratively in a studio environment to create a shared

understanding of the people they design for (and with) and the product ideas they develop. Topics covered include design thinking, ethnographic methods, concept development and interaction design. (Olin College 2019a)

Other distinct differences at Olin College include a lack of academic departments, a 50% tuition scholarship for all students, an extended on-campus interview before admission (to see if they can collaborate and work in teams), and a simplified list of possible degrees that includes only three majors—all in engineering. Unlike many engineering programs, Olin is gender balanced. All of the Olin students participate in a year-long design Capstone Project with companies and other institutions, or an Affordable Design and Entrepreneurship Project, often at international locations to help "communities around the world to address challenges endemic to poverty and democratize opportunity." Olin's partnerships with nearby Babson College, Brandeis University, and Wellesley College have also broadened the courses available to Olin students into business and arts and humanities (Olin College 2019b).

Within Olin College is a new form of engineering education that creates a palpable sense of excitement. This was apparent in our visit to campus, as students showed us the robotics lab, the machine shops, newly built aerodynamic pedal cars, motor-driven dune buggies, and even new furniture for common areas (Fig. 4.3). Just as in the pioneering "partner year" of 2001 when the first students were testing concepts with faculty, Olin in 2018 is developing new ideas with its students and having a lasting impact in engineering education.

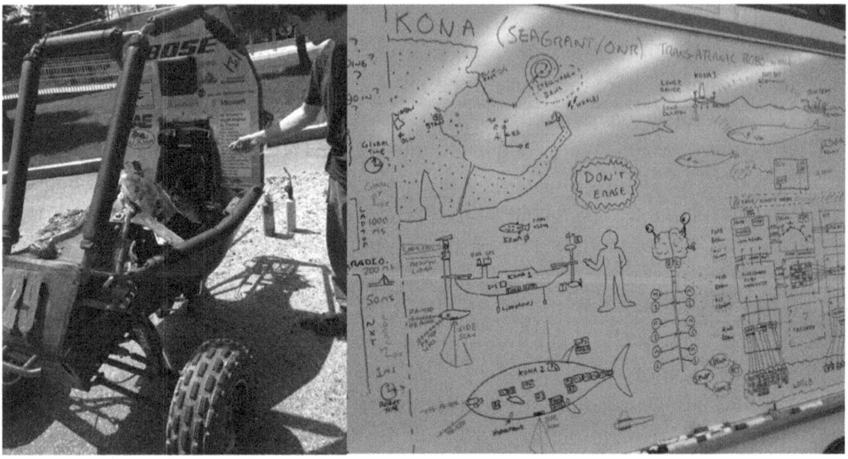

Fig. 4.3 (Left) An Olin College student takes a break from welding and shows us the dune buggy he and his team have built from scratch, which they use in intercollegiate "challenge" races. (Right) A whiteboard in the robotics lab, where students are constructing a "transatlantic robo-whale," robotic fish, and even a robotic kayak

4.4 Liberal Arts Engineering

The emergence of new programs and centers for more human-centered and design thinking types of engineering education is well illustrated in the cases of Yale Engineering, Cal Poly and Olin College. As awareness grows about the vital importance of the interplay between social sciences, humanities and the arts and engineering, new types of curricula are being developed that are often described as "Liberal Arts Engineering." Citing a wide range of sources, including surveys from employers, educational research, and a cross-sectional survey of curriculum within the US, a recent National Academies report advocated adoption of an integrative model of engineering and medical education that:

> Intentionally seeks to bridge the knowledge, modes of inquiry, and pedagogies from multiple disciplines—the humanities, arts, sciences, engineering, technology, mathematics, and medicine—within the context of a single course or program of study. In such a model, professors help students to make the connections between these disciplines in an effort to enrich and improve learning. (National Academies 2018)

The report also noted that broader and more integrative training develops "increased empathy, resilience and teamwork; improved visual diagnostic skills; increased tolerance for ambiguity; and increased interest in communication skills." Other studies have highlighted the centrality of communication and interpersonal skills in a wide range of STEM fields. One example was from a survey of MIT mechanical engineering alumni from the classes of 1992–1996, 2003–2007 and 2009–2013. The alumni were asked to rank the usefulness of a wide range of skills in their careers in engineering. Ranked among the top five were "Communication, Personal Skills, Independent Thinking, Teamwork, and Professional Attributes." At the bottom of the list of 25 skills were content-based technical skills such as "Mechanical Behavior of Materials, System Dynamics, Heat Transfer, Fluid Mechanics, and Thermodynamics"—all of which were traditionally the mainstays of the traditional engineering curriculum (Wang 2015). Other research studies have shown that traditional STEM pedagogies discourage women and people of color from persisting in STEM fields compared to broader and more integrative pedagogy (Byars-Winston et al. 2010; PCAST 2012).

Benefits from diversity within teams of engineers include *cognitive diversity* which promotes better problem solving and enhances learning. As the report notes:

> Diversity in higher education encourages critical thinking, challenges practitioners to consider multiple points of view; promotes teamwork across disciplinary boundaries; encourages an appreciation of differences in background, training, and aptitudes; and highlights the importance of the contexts in which education, work, and citizenship develop and are displayed. (National Academies 2018)

The National Academies report provided several examples of curriculum and courses within engineering and medicine that successfully integrate humanities and arts within their programs. Courses for STEM majors were divided into categories based on the type of integration—within courses, between courses in a curriculum and between curriculum and co-curriculum. Courses with "in-course integration" might bring a creative or arts assignment into a STEM course, or add guest lectures or reading in humanities and arts. Examples include an undergraduate neuroscience course with a film-making assignment, adding a haiku writing project in a STEM course, building sculptures based on 3-D shapes of biomolecules, and adding a history component to a materials science course. Each of these courses demonstrated greater engagement, increased communication skills for the students, and exploration of STEM subjects at higher levels in Bloom's taxonomy. Courses with integration across courses (such as linking engineering with graphics or writing courses) were also studied. Examples include linked engineering courses with courses in graphic design and ethics, a first-year engineering program highlighting connections between engineering and humanities, and an Imaging Science degree program integrated with a BFA in Film and Animation. These courses were found to provide increased graduation and retention rates, higher percentages of students obtaining jobs, and higher GPAs among students in these programs, compared to more traditional courses.

With funding from the National Science Foundation, a new *Liberal Studies in Engineering* program was developed by a group headed by Louis Bucciarelli, former MIT mechanical engineering professor, and David Drew, former Dean of the Claremont Graduate University education school. The Liberal Studies in Engineering group met at the National Academy of Engineering in 2015 to discuss a new degree program that would infuse liberal arts into an engineering degree. The meeting included Deans and Directors of programs at leading engineering institutions such as Purdue, WPI, MIT, Harvard, Vassar, Brown, Cal Poly, Dartmouth, USC, UC Berkeley, RIT and many others (Bucciarelli et al. 2015). Many of the representatives also came from international programs, such as engineering universities in the Netherlands, France, Denmark, as well as from the international liberal arts program in Singapore at Yale-NUS College.

The meeting discussed new pedagogical paradigms for engineering, and explored more interdisciplinary approaches, as well as new temporal structures within the curriculum such as short online courses, specialized mini-modules and intensive experiential learning. The meeting discussed how to transcend silos within colleges and universities to facilitate the needed innovations and collaborations to implement these innovations. Several examples of innovation and inclusion were shared which arose from what amount to "vacant lots" within an institution—these include maker spaces, modules within courses, summer programs, online efforts, gap year programs, internships and first-year interdisciplinary seminars. All of these spaces exist outside the existing highly regimented disciplinary boundaries of departmental semester offerings, and the trick is to transfer the lessons learned from these laboratories into the regular curriculum to transform engineering education.

In an opening session, the problem with engineering education was explained by David Drew, with a story from Chaim Weizmann about a sea voyage with Albert Einstein in the 1920s. In the story Wiezmann described how he had dinner with Einstein every day, and "Einstein explained his theory to me every day, and on my arrival I was fully convinced that ***he*** understood it" (Schwinger 1986). This story also illustrated how a lecture-based form of engineering education tends to be more of a demonstration of the professor's understanding than an environment that promotes student learning. Drew also noted that the disconnect between Engineering and other disciplines prevents communication. He invoked a quote from C.P. Snow's famous lecture on "The Two Cultures":

> There have been plenty of days when I have spent the working hours with scientists and then gone off at night with some literary colleagues. … I got occupied with the problem of what, long before I put it on paper, I christened to myself as the 'two cultures'. For constantly I felt I was moving among two groups—comparable in intelligence, identical in race, not grossly different in social origin, earning about the same incomes, who had almost ceased to communicate at all, who in intellectual, moral and psychological climate had so little in common that instead of going from Burlington House or South Kensington to Chelsea, one might have crossed an ocean. (Snow 1959)

Many new engineering courses and programs were shared during the meeting to remedy the problem, by blending diverse disciplinary and cultural perspectives with intense experiential learning. Three illustrative examples of new types of engineering curriculum included courses from the D-Lab at MIT, the Global Problems Seminar at WPI, and the Liberal Arts and Engineering Studies program at Cal Poly.

The D-Lab at MIT was developed by MIT professor and 2004 MacArthur Fellow Amy Smith. The D-Lab has created over 20 MIT courses over the past 15 years that blend service learning and rigorous engineering to develop practical solutions to global poverty challenges (MIT D-Lab 2019a). The program began with the "Haiti Class" in 2002 which enabled students to help develop a system for converting unused sugar cane fiber into useful fuel for villagers. Since 2002, the D-Lab has grown to dozens of courses reaching over 2000 students, who connect with communities in Botswana, Brazil, Colombia, El Salvador, Ghana, Guatemala, India, Indonesia, Haiti, Ghana, Mali, Morocco, Nicaragua, Pakistan, Peru, Philippines, Tanzania, Uganda, Zambia, and other countries. The D-Lab courses are aligned with three "pillars"—Education, Research and Innovation practice. The courses practice inclusive design pedagogy, and work with local communities to develop solutions that are practical and have thoroughly considered societal impacts. The community of students and partners meet regularly in a series of International Development Design Summits, which currently includes over 24 summits in 11 different countries (MIT D-Lab 2019b).

Another compelling engineering curriculum is the Great Problems Seminar from Worcester Polytechnic Institute (WPI) where first-year students immerse into a project-based curriculum and help solve problems of global importance. One component of this program is a Global Projects program, which is an engineering-infused international experience where students travel to one of WPI's 40 + project centers across the world, where WPI has ongoing collaborations with communities Namibia, Paraguay, and Costa Rica and dozens of other countries (WPI 2019). Some of the seminars are based on themes such as "Heal the World," which explores global medical challenges, "Shelter the World," which seeks solutions for affordable housing across the globe, and "Feed the World" which looks at how to improve the systems for growing, harvesting and distributing food in local and global contexts. The Great Problems seminars also include the Humanitarian Engineering curriculum

(Boudreau et al. 2016), where students are able to blend methodologies of engineering and humanities to immerse in complex problems from recent history. Students use historical sources and work in teams to research and present real problems, including role-playing exercises, where different groups of students take on the roles of diverse and opposing stakeholders. One example is solving the sanitation and water access in Worchester, Massachusetts in the 1890s, which includes a complex mix of problems with labor rights and social inequality as well as technical challenges.

Cal Poly provides a third example of innovative engineering education in its Liberal Arts and Engineering Studies curriculum (Cal Poly LAES 2019). The program blends computer graphics, electrical and manufacturing engineering with liberal arts topics such as cinema, theatre and studies of culture, society and technology. The program trains students for careers in audio engineering, digital media production, game design, and digital publishing, as well as in many educational and policy making roles requiring advanced communication skills and technical proficiency with emerging technologies. The LAES program works closely with Cal Poly's Center for Expressive Technologies to develop "interdisciplinary projects focused on the intersection of human expression and technology." Cal Poly is creating a new building that will feature a set of "Expressive Technology Studios" for creating music, soundscapes, digital effects, and other expressive technologies, as well as offering flexible collaborative space for the LAES students (Cal Poly CET 2019).

Since the conference on Liberal Arts Engineering, Drs. Buccharielli and Drew have written a report on their findings. The report describes additional exemplary courses that expand the boundaries of engineering to encompass Intersections of Technology, Social Justice and Technology—the name of a course at Wellesley College, and how we might approach Engineering Peace—the name of a course at San Diego State University. The report also notes that many first-year or introductory courses in such as Lafayette College's Integrated First Year Experience, and Smith College's Engineering for Everyone have made excellent progress engaging large number of students in the intersections of engineering, humanities and social sciences. The report also notes, however, that institutional barriers block progress in many universities through a lack of incentive structures for faculty to participate in interdisciplinary courses and resistance from both faculty and administrators to courses that blend expertise within and outside of engineering.

The most hopeful and effective elements within the new era of engineering include not only the courses and curricula described above, but design studios, student design competitions, and programs that have blended expertise in social science and engineering, particularly with collaborations between anthropology and engineering. New maker and hacker spaces like the Yale CEID, and within communities are diversifying engineering and engaging (Riley et al. 2017). Using ethnographic methods and Critical Discourse Analysis (CDA), Riley documented how Maker and Hacker spaces in California, New York, Colorado and Virginia, have stimulated innovative design thinking and have developed inclusive practices that has been effective fostering engineering skills across diverse populations enabling enhanced diversity by gender, socioeconomic level and race.

New curricula that enhance connections between engineering and the arts, psychology, and economics offer promise to give students vistas of the many ways that the complexity of human experience affect the design and impact of new technologies. From this effort can come a more humanistic form of engineering education that will have profound impact in helping students solve the many complex and socially rooted problems facing society in the coming decades.

References

Boudreau, K., Hanlan, L. R., Dodson, L., DiBiasio, D., Abel, C., Sullivan, J. M., Gaudette, G. R., Bergendahl, J., & Demetry, C. (2016). *Humanitarian engineering, past and present: A role-playing first-year course*. OEC: Online Ethics Center for Engineering and Science. Retrieved from https://digitalcommons.wpi.edu/gps-research/10.

Bucciarelli, L., Drew, D., & Tobias, S. (2015). *Liberal studies in engineering—Workshop report*. Retrieved January 8, 2019, from https://core.ac.uk/download/pdf/78062001.pdf.

Byars-Winston, A., Estrada, Y., Howard, C., Davis, D., & Zalapa, J. (2010). Influence of social cognitive and ethnic variables on academic goals of underrepresented students in science and engineering: A multiple-groups analysis. *Journal of Counseling Psychology, 57*(2), 205.

Cal Poly CESAME. (2019). *Learn by doing lab*. Retrieved January 8, 2019, from https://cesame.calpoly.edu/content/learn-by-doing-lab.

Cal Poly CET. (2019). *Center for expressive technology*. Retrieved January 8, 2019, from https://cet.calpoly.edu/about.

Cal Poly CIE. (2019). *Revolutionizing entrepreurship*. Retrieved January 8, 2019, from https://cie.calpoly.edu/.

Cal Poly EOP. (2019). What is SI? Retrieved January 8, 2019, from https://eop.calpoly.edu/si/what-is-si.

Cal Poly LAES. (2019). *Freedom to choose*. LAES Brochure. Retrieved January 8, 2019, from https://content-calpoly-edu.s3.amazonaws.com/laes/1/documents/LAES_Trifold_Brochure.pdf.

California Polytechnic State University San Luis Obispo EPIC. (2019). *Engineering Possibilities in College (EPIC)*. Retrieved January 8, 2019, from https://epic.calpoly.edu/draft-about/.

Clark, L. (2018, April 19). Cal Poly is the Whitest Public University in California—By a lot. *San Luis Obispo Tribune*. Retrieved January 8, 2019, from https://www.sanluisobispo.com/news/local/education/article209195019.html.

Cubesat.org. (2019). *Origin of the new space revolution*. Retrieved January 8, 2019, from http://www.cubesat.org/.

Cunningham, W. J. (1992). *Engineering at Yale: School, Department, Council, 1932–1982* (xiii, 217 p.). Transactions/The Connecticut Academy of Arts and Sciences. New Haven, CT: The Connecticut Academy of Arts and Sciences.

Guizzo, E. (2006, May 1). The Olin experiment. *IEEE Spectrum*. Retrieved January 8, 2019, from https://spectrum.ieee.org/at-work/education/the-olin-experiment.

Harvard Magazine. (2011). *'Hi' to the Harvard Innovation Lab*. Retrieved January 8, 2019, from https://harvardmagazine.com/2011/11/harvard-innovation-lab-opens.

Kembel, G. (2007). *d.school operators handbook*. Retrieved January 8, 2019, from https://dschool.stanford.edu/resources/george-kembels-dschool-operators-handbook.

Letchford, J. (2009). A small school with grand ambitions. *Yale Daily News*. Retrieved January 8, 2019, from http://yaledailynews.com/blog/2009/09/22/a-small-school-with-grand-ambitions/.

Marcus, J. (2002, Spring). An unknown quantity. *National Cross Talk*. Retrieved January 8, 2019, from http://www.highereducation.org/crosstalk/ct0202/news0602-unknownq.shtml.

MIT D-Lab. (2019a). *Amy Smith*. Retrieved January 8, 2019, from https://d-lab.mit.edu/about/people/amy-smith.

MIT D-Lab. (2019b). *International design summits*. Retrieved January 8, 2019, from https://d-lab.mit.edu/about/international-development-design-summits.

National Academies of Sciences, Engineering, and Medicine. (2018). *The integration of the humanities and arts with sciences, engineering, and medicine in higher education: Branches from the same tree*. Washington, DC: The National Academies Press. Retrieved from https://doi.org/10.17226/24988.

Olin College. (2019a). *User oriented collaborative design*. Retrieved January 8, 2019, from http://www.olin.edu/academic-life/experience/user-oriented-collaborative-design/.

Olin College. (2019b). *The academic experience*. Retrieved January 8, 2019, from http://www.olin.edu/academic-life/experience/.

PCAST. (2012). *Engage to excel: Producing one million additional college graduates with degrees in science, technology, engineering and mathematics*. Retrieved from https://eric.ed.gov/?id=ED541511.

Riley, D. M., McNair, L. D., & Masters, S. MAKER: An ethnography of maker and hacker spaces achieving diverse participation. Retrieved January 8, 2019, from https://vtechworks.lib.vt.edu/handle/10919/82443.

Schwartz, J. (2007, September 30). Re-engineering engineering. *New York Times Magazine*. Retrieved January 8, 2019, from http://www.nytimes.com/2007/09/30/magazine/30OLIN-t.html.

Schwinger, J. (1986). *Einstein's legacy—The unity of space and time*. New York: Dover.

Snow, C. P. (1959). *The Rede Lecture, 1959—The two cultures*. Retrieved January 8, 2019, from http://s-f-walker.org.uk/pubsebooks/2cultures/Rede-lecture-2-cultures.pdf.

Vanderlick, K. (2011). *Message from the Dean*. Yale Engineering 2011. Retrieved January 8, 2019, from https://seas.yale.edu/sites/default/files/Yale_SEAS_2011_Publication.pdf.

Wang, K. (2015). *Study on the careers of MIT Mechanical Engineering undergraduate alumni*. Doctoral dissertation, Massachusetts Institute of Technology.

Werner, D. (2018). Cubesat co-inventor Jordi Puig-Suari sails into the sunset. *Space News*, August 7, 2019. Retrieved January 8, 2019, from https://spacenews.com/cubesat-co-inventor-jordi-puig-suari-sails-into-the-sunset/.

WPI. (2019). *Great problems seminar*. Retrieved January 8, 2019, from https://www.wpi.edu/academics/undergraduate/great-problems-seminar.

Yale SEAS. (2018). *Yale engineering*. Retrieved January 8, 2019, from http://www.yaleseas.com/magazine/issues/2017–18/index.html.

Yale SEAS. (2019). *Y-Shaped engineers*. Retrieved January 8, 2019, from http://seas.yale.edu/about/y-shaped-engineers.

Chapter 5
Online Education in STEM

Abstract From a series of visits to the leading online learning centers during the "year of the MOOC" in 2013, the roots of the online learning phenomenon are explored, and their applications to STEM disciplines are described. Site visits to the headquarters of Coursera, edX, Udacity, and HarvardX are combined with an overview of their history and current programs, including interviews with the founders and leaders of all of these major online education centers. The evolution of online learning from its early days into the present is described, along with new types of online learning that include "stackable micro-credentials," hybrid courses, and online degree programs. The future of online learning and its impacts on colleges and universities is also described, with examples of possible new types of online learning environments in STEM fields.

Keywords Online learning · Hybrid learning · Micro-credentials · MOOC · Online education

5.1 From MOOCs to Microcredentials

In 2013 we experienced "The Year of the MOOC"—the Massively Online Open Course. In some ways the onset of online learning was reminiscent of the discovery of Dark Energy. Both Dark Energy and MOOCs were unknown a decade ago, and both threatened to dominate and expand exponentially and determine the fate of the physical universe and the universe of higher education. The headlines in both academic publications and the popular press were describing the period as the era of MOOCs, and forecasting the extinction of the university as we know it.

The development of online education was a dramatic advance that enabled literally hundreds of thousands of students to take a single online course, such as the more than 150,000 students who took Sebastian Thrun's course "Introduction to Artificial Intelligence" in 2011. The technology promised to not only reach unprecedented numbers of students, but also to have a breathtaking geographic reach where students in courses hailed from Latvia to Lybia, from Yemen to yak herders in Mongolia! It appeared that the arrival of the MOOC was historic as a "transformative" technology for higher education. But now several years after—what does online

B. E. Penprase, *STEM Education for the 21st Century*,
https://doi.org/10.1007/978-3-030-41633-1_5

education mean for universities, liberal arts colleges, and students—and where is the technology going? Can professors adjust to this "revolution" or should they? And what will this explosion of online education technology mean for the quality of teaching in universities and colleges—will it improve teaching and learning? or replace professors with recorded lecturers from other institutions in the new "flipped" classroom environment?

During my ACE fellowship in 2012–13, I had the unique opportunity to explore the world of online education during the height of the MOOC hysteria. The year included visits and interviews with the founders of the main online learning centers—Coursera, Udacity, edX, HarvardX, and Stanford's online education center. From these visits we were able to both learn about the nature of the technology being developed at these online learning sites, but also something of the institutional cultures of these innovative companies, and the possible effects of online learning on today's universities and colleges. I describe both the nature of these online education companies and their courses as they appeared in 2012–13, and five years later in 2018 after online courses had a chance to be tested and some of the limitations of the early versions of MOOCs became more apparent.

5.2 A Brief History of MOOCs and Online Learning

Many of the main MOOC providers are only a few years old, but they arise from over a decade of increases in the development of online content and the rise of the learning management system (LMS) in education. The earliest forms of online education include the PLATO (Programmed Logic for Automated Teaching Operations) system from 1960, developed at the University of Illinois, which was a completely online system for experimenting with communications between teachers and students, much like a text-based chat system. By 1969, the Arpanet—the ancestor of our modern internet—was founded by the Department of Defense, along with some of the first protocols for email and online discussion forums. By the early 1970s, experiments with video recordings and distance learning were conducted that had some of the feel of modern MOOCs. Fully online courses in logic and set theory were offered at Stanford University from 1972 to 1992, and further experiments in distance learning in the 1970s included community college and local television stations collaborating to combine video with phone-based communications between distributed groups of students and instructors. Several of Stanford's engineering courses were offered remotely via close-circuit television to remote sites in Silicon Valley, and these sort of video courses proliferated across the country during the 1980 and 1990s, and had many of the features of a modern MOOC. Also during the1980s the first LMS or Virtual Learning Environment (VLE) systems were developed, which were the precursors of modern systems like Moodle or Blackboard.

The development of online materials through web sites progressed steadily during the 1990s and the early 2000s, to a point where nearly every college instructor was expected to develop a class web site, providing syllabus and assignments through an online portal, and in many cases interactive features such as blogs, chat rooms, and wiki pages were provided for students through systems such as Blackboard, Moodle, Sakai, classesv2 Canvas and other LMS environments.

One interesting strand in the evolutionary development of online learning was based on the concept of "connectivism" (Siemens 2005), in which a networked community works together with a shared library of "digital artifacts." This concept developed into one of the first online courses in 2008, entitled *Connectivism and Connected Knowledge* (Downes 2008), which emphasized an active role of learners in co-creating knowledge in an online community. This more interactive form of online course was known as the connectivist MOOC, or cMOOC, and was adapted into a course entitled *Change11* that included several thousand students, and provided the basis for one of the earliest studies of online learning (Milligan et al. 2013).

One of the early systems for offering online lectures was iTunesU, which until recently offered thousands of courses, with over 600,000 free lectures. Yale University was an early leader in online courses with its OpenYale initiative, which provided over 30 Yale courses free of charge with recordings of full lectures and a web site offering course materials. Early experiments such as the MIT Open CourseWare initiative expanded the range of free content available online to include huge libraries of interactive physics and engineering simulations, assignments and quizzes, and recorded lectures. These resources evolved into the modern edX consortium, which originally was a collaboration between MIT and Harvard to offer a unified platform for online courses (Fig. 5.1).

One of the first complete online courses which would qualify as a MOOC was the 2011 course in Machine Learning from Sebastian Thrun at Stanford, which became famous after enrolling over 80,000 students. Other Stanford professors also offered very popular online courses in computer science, such as courses in programming and artificial intelligence from Andrew Ng and Daphne Koller. The success of these courses, and the enormous numbers of students who enrolled and completed the course, give them the title of MOOC (Massively Open Online Course). These three instructors later became the founders of two of the leading online education centers, Coursera (Koller and Ng) and Udacity (Thrun). The diagram below shows a timeline and evolutionary tree of the online education providers from the early days to the present.

The MOOC or online course is perhaps the limiting case of technology-enhanced teaching, which can span from a conventional course supplemented by a web site or LMC, to a hybrid course that might include some online assessment and videos. For many instructors, the addition of these technologies are natural outgrowths of the growing role of computers in our daily lives. In some cases, they have become the default learning environment in today's colleges and universities. Transitioning from a hybrid course toward a fully online course requires an extra layer of technology to seamlessly integrate content and assessment in an engaging interface that will retain students and guide their progression through the material, with some amount of assessment, tutoring and support.

A schematic diagram that portrays this range of learning technologies can be constructed using a series of concentric rings (Fig. 5.2). The outermost rings, which

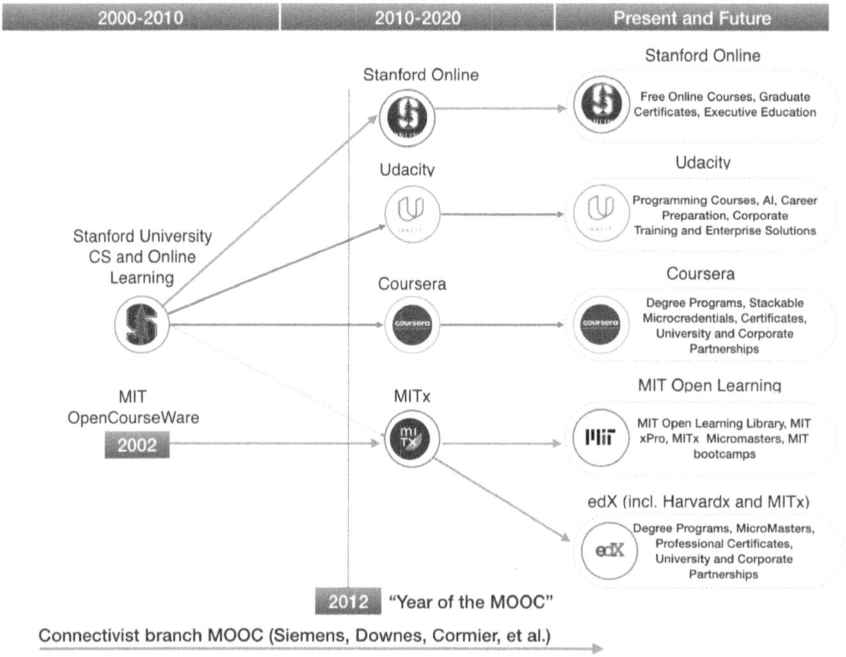

Fig. 5.1 Evolutionary tree of the MOOC providers of today (original figure based on Hill and Feldstein 2013)

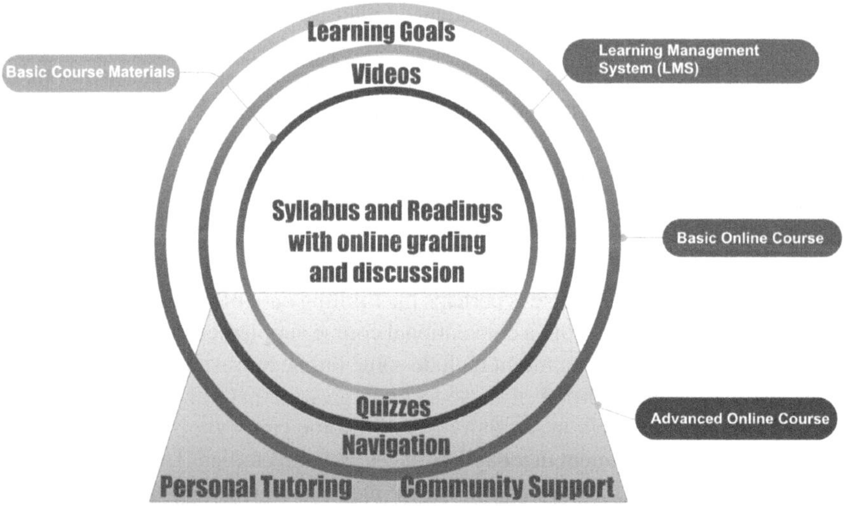

Fig. 5.2 Diagram showing concentric rings showing the expansion of a course from traditional materials to a full online course (original figure based on Hill and Feldstein 2013)

include systems to unify course content, assessment, and media with an "intelligent tutoring system" and "community discussion support" are what typically makes the difference between an LMS and an online course (Hill and Feldstein 2013).

This outermost layer unifies all of the class content—lectures, assignments, discussion forum, and quizzes—in a single interface. In many cases these interfaces will also include interactive chat (video or text), game-like simulations and scenarios, and systems for uploading essays and other creative work for grading or sharing with the community. The development and evolution of these interfaces is what is new, and the pace of development has been dizzying, and will continue to evolve in the coming years.

To gather some first-hand information about MOOCs, their creators, and the vision of the innovators and designers of these systems, I visited a number of the leading centers for online learning to hear from their founders and to see their technology as it is being developed, as demonstrated by their technical staff back in 2012 and 2013. The first visit was to the Silicon Valley group of online education centers that included meetings at Stanford, Coursera and Udacity. We felt we were getting a glimpse of the future of higher education, and also of new forms of institutional culture. A month later I visited the headquarters of edX, in Boston, talked with some of their staff, and met with the edX CEO and founder, Anant Agarwal. Further visits to Harvard's Bok Center for Teaching and Learning enabled me to see how HarvardX and online education were being implemented and the dramatic impact HarvardX was having on campus culture.

5.3 Coursera

The largest, and perhaps most famous of online course providers, is Coursera. Coursera was founded in April 2012 by Daphne Koller and Andrew Ng, who both put their careers at Stanford on hold to work full time on Coursera. Koller and Ng had worked together since their days as graduate students at UC Berkeley, and the Coursera project arose naturally from their work at Stanford. Koller was working with "flipped classes" for years before Coursera, and Ng had a course in machine learning that enrolled 104,000 people in 2012. They joined forces to build new software and an organization for coordinating the activities of the hundreds of thousands of students and thousands of professors who together offer courses under the Coursera organization. The goal, according to Koller, was "to bring the best quality education to as many people as we could." Ng and Koller hope that by bringing courses to every corner of the world, it would enable hidden talent to be found to trigger what Koller describes as a wave of innovation:

> this would enable a wave of innovation, because amazing talent can be found anywhere. Maybe the next Albert Einstein or the next Steve Jobs is living somewhere in a remote village in Africa. And if we could offer that person an education, they would be able to come up with the next big idea and make the world a better place for all of us. (Koller 2012)

The Coursera organization was startling in its early successes and raised over \$65 million dollars in its first year, from an early investment in April 2012 of \$16 million from John Doerr at Kleiner Perkins and Scott Sandell at NEA, joined with a \$43 million investment from a consortium that includes GSV capital, International Finance Corporation (IFC), representing the World Bank's investments, Laureate Education, Inc., and Learn Capital. These last investors chose Coursera both for its reputation in education technology circles, and its connections with emerging markets. The funding was intended to enable Coursera to expand its efforts to reach the developing world, and to target under-served populations globally, such as young women (Farr 2013). A further boost to Coursera was provided when the American Council on Education recommended five Coursera courses for college credit in January 2013. These recommended courses include Algebra and Pre-Calculus from UC Irvine, Intro to Genetics and Evolution from Duke, Bioelectricity from Duke and a Calculus course from the University of Pennsylvania (Kolowich 2013). By October 2013 Coursera grew to include over 5.2 million students, and 107 university partners in 20 countries. By 2017, these numbers only continued to rise, with over 30 million students, and over 2700 active courses (Class Central 2017). The achievements of Ng and Koller were widely recognized, as their company was awarded the "Best Startup of 2012" by TechCrunch in January 2013, and both Ng and Koller were awarded named as two of the Time 100 most influential people in the world in 2013.

All of this growth and success is not without some controversy, however. The rush to build online courses alienated many faculty, and caused some to worry about pernicious effects of "corporate style" education that could threaten faculty jobs. An article in the Nation described the phenomenon as "the Coursera Hype Machine":

> The hype, and the money behind it, are creating new pressures on colleges and universities to replace faculty members in the classroom with online video lectures. That, they say, would bring the genius of the free market to higher education by reducing labor costs (i.e., professors) and introducing economies of scale (teaching tens of thousands of students in a single course). (Wiener 2013)

Both politicians and corporate financiers were eager to get involved in the world of MOOCs. The University California Board of Regents was urged by California governor Jerry Brown to move quickly to award credit for over-enrolled courses, and a bill introduced in February 2013 to the California State Senate would have required colleges and universities to award credit in the fifty most over-subscribed introductory courses. Darrell Steinberg, sponsor of the bill in the Senate, described why this was important:

> We want to be the first state in the nation to make this promise: No college student in California will be denied the right to move through their education because they couldn't get a seat in the course they needed. (Lewin 2013).

The UC faculty were instantly galvanized to oppose this measure and were able to mobilize opinion and votes to help defeat the bill, which was re-worded into a much weaker bill that merely recommended that faculty accept online courses for credit. The surge of interest from the corporate and foundation sector in MOOCs trigged a backlash of fear and resentment in many faculties at Colleges and Universities, but the online education sector continued to grow. It was in this turbulent time that we visited Coursera's headquarters in Mountain View California, during Spring of 2013.

5.4 Coursera HQ Visit in the Year of the MOOC

In 2013, Coursera occupied a modest office suite tucked away in the second floor of a nondescript professional building in Mountain View, CA. Coursera's business partnership director, Chris Heather, greeted our party of ACE fellows warmly when we arrived and he and Diane Lee, course operations director, gave us an overview of their operations in a small boardroom just the right size for our 12 ACE fellows (Fig. 5.3). It seemed mildly ironic that this world-changing organization reaching hundreds of thousands of students could fit its entire operation within the office suite size of an average medical practice, with a board room that only held about 12 people! After a while Andrew Ng, co-founder of Coursera, came in. Chris and Diane handed over the computer to Andrew, and we had a chance to get an overview of MOOCs and Coursera from one of the leading figures in the world in intelligent machine design.

Andrew described the platform and the philosophy of Coursera, and its role in offering access to higher education across the earth independent of location, wealth, or status. The scale and potential impact of the new technology was stunning. Part of the challenge of a MOOC which I had not realized was identifying precisely *who* is taking a course. Getting this right is crucial for awarding credit for these courses, but since people never meet each other, sophisticated biometric identifications are used to verify the identify of the MOOC students. For a fee (typically about $50), the software was designed verify identity both by video which is used each time an

Fig. 5.3 Diane Lee, one of our gracious hosts from Coursera, poses by a world map highlighting all the institutional partners of Coursera circa 2013—at this time Coursera had 70 partners from around the world, and was hosting courses taken by 3.7 million students

assignment is submitted, by an analysis of typing patterns, and from a verification of identity through a passport of driver's license. Another service called e-Proctor was used to watch students take exams.

The other interesting thing I had not thought about is how instructors grade assignments from the tens of thousands of students often enrolled in a MOOC. The Coursera platform in the initial versions employed "crowd-sourcing" and peer grading as the solution. Andrew presented results from classroom studies that showed how students' self-evaluations and peer evaluations correlate very well with instructor grades from a wide range of courses. Each student was asked to grade five others' papers, and in return got feedback from five peers who are also taking the course. The software monitored the reported scores from the assignments, looked for outlying and rejected points which could represent bad grading, and provided a robust system of peer grading with aggregated grades from students within the course.

Coursera's initial platform offered "synchronous" courses where students began on a fixed date, and proceeded at a measured pace through the material with regular deadlines and some of the other features of an old-style university course. Andrew was emphatic that this synchronicity enhances learning, since the community of students experience the course together, and can post questions and help each other through meet-ups during the semester. He even had a plot of web traffic that showed a "heartbeat" in time which was a series of regular and strong upticks in the traffic right before an assignment was due! In addition to the legions of students, there also is a rapidly growing world-wide community of instructors building courses on Coursera. These instructors are linked by online forums, and in-person annual meetings, and form something of a meta-global university faculty, who are linked not by geography or institutional culture but by online platform and technology. This is a fascinating and overlooked byproduct of the MOOC movement, since the professors, as well as students, experience a new global community to learn from.

Some instructors arranged for "Google Hangouts" to stage online discussions with students. These were videoconferences that can be recorded and offered as material for other students to view (only about 20 can be in a hangout at once) and provided some level of face time with students. The instructor of a Coursera course often could receive 100,000 emails directly, and in some cases would use former Coursera students as experts on the online forums and to provide help to other students. Coursera built its institutional partnerships with the understanding that each institution would cover the costs of producing the course, and for paying the instructor and any necessary overhead for developing the course on campus. Coursera provided the platform, limited online support, and organized the schedule for the course, which is locked into a semester-like timeline. Coursera also emailed students while they work, reminding them of deadlines and encouraging them to complete the course.

One development after 2013 was the growth of online degree programs, specializations and micro-credentials. By 2017, Coursera offered 257 different credentials and specializations, and complete online degree programs that include a Bachelor of Science in Computer Science from University of London, Master of Computer Science from Arizona State and University of Illinois, and other degrees such as an MBA and Master of Science in Accounting from University of Illinois, and new masters programs in Public Health. All of these programs are fully accredited, and in most cases cost a fraction of the comparable on-campus degree program.

Coursera also found many ways to "monetize" the online course environment and to reach new markets of students. In 2016, Coursera began to require payment for assignments to be graded in some online assignments (Shah 2015), and required additional fees for specializations, which were three linked courses combined with a capstone project. The number of specialization and micro-degrees continue to increase, and are an increasingly popular option for working professionals upgrading their skills. The shift in emphasis for Coursera has moved the platform into new territory where the online courses are not replicating a mass-audience university lecture, but providing entirely new options for employees at companies and for "lifelong learners." Coursera is expanding its online degrees as well, with 15–20 new online degree programs expected by 2019 (Shah 2017a). New degrees announced in 2018 include a Master of Computer Science from Arizona State University, a Global Master of Public Health from Imperial College London, a Master of Computer Science from the University of Illinois, a Bachelor of Science in Computer Science from the University of London, a Master of Applied Data Science from the University of Michigan, and a Master of Public Health from the University of Michigan (Sinha 2018).

Rather than replicating or disrupting the university, Coursera is creating new formats and markets for higher education, with tie-ins to companies through its Coursera for Business arm, which allows companies to offer their employees access to all of Coursera's courses, including specialized tools for community and monitoring progress, for a small fee per employee. This extension of the higher education space allows Coursera to reach learners who otherwise would not be able to access needed courses, and provides an interesting counterpoint to on-campus traditional higher education. As such Coursera is able to reach entirely different sectors of students—full-time workers in companies, people with families looking for a second degree, and what otherwise might be considered "non-traditional" students. By bringing the high-quality intellectual content to these students, it is able to serve the broader societal mission of the various universities who set up the courses in the first place. These new online degree programs also have enabled Coursera to differentiate its platform and to add new capabilities. These the new platform for corporate training, which includes a local scheduling and analytic program as well as communication capabilities within the company, and a more advanced and interactive platform for online degree programs that includes many tools for advising, for group discussions and for the synchronous social interactions that previously were only available for on-campus courses.

As Nikhil Sinha, former Chief Content Officer of Coursera, described in a 2018 visit to Silicon Valley, "Coursera is only six years old. I like to think of the first three years as bringing the university to the world in time and in space. The second three years is Coursera gathering all of the materials and things they have learned, and bringing those back to the campuses." Through the use of the new platforms, Nikhil reports that many of the partner institutions are now able to meet the needs of their on-campus students in new ways. Examples include engineering students at University of Illinois who are now able to enroll in business courses through the University of Illinois Coursera online degree program, and schools like University of Illinois and University of Michigan that now can easily work together to offer courses and combining expertise from their campuses in new ways. Nikhil also pointed out that some of the earlier criticisms of MOOCs from 2013 were premature, both as the technology was just getting started, and as we were less aware of the types of students who would be most interested in online courses. For many students learning computer science, or for those with busy lives and families, the lack of the social environment and synchronous component in no way detracts from learning but in fact makes the learning more appealing than an on-campus course. By reaching a large segment of a company employee base, the Coursera corporate partners are able to upgrade the skills of a huge number of workers who would be out of reach of existing executive training programs.

Research from over five years of online courses has shown that 85% or more of Coursera's learners are 23 or over. This means that the online sector, rather than being a competitor or "disruptor" of campus education at universities and colleges, provides a complementary role in reaching "life-long learners" and also workers who need to advance their careers through online courses and credentials. Rick Levin, the former CEO of Coursera, stressed in a 2018 interview the ways in which online learning at Coursera has evolved since the 2012 "Year of the MOOC." For many of the learners using Coursera courses, "offline is not an alternative" to the online platform. Levin stressed that Coursera can provide opportunities for both life-long learners, and for career advancement and practical skills in fields like data science for career learners. New programs like "verticals" in Public Health and online Master's programs are further advancing the chances for career building through Coursera. Coursera is also partnering with leading universities with credentialed degrees, such as programs with Imperial College, London, University College London, and University of Illinois and Michigan. New students interested in these programs, who may not have been in College for years, can begin by sampling some online courses from Coursera to "prove themselves" and then enroll with a greater assurance of success in the online degree programs. Levin pointed out how the online courses are creating a new and more global community of students and faculty, with strong interest within European companies for the online training, and collaborations with world universities in the UK, USA and Japan creating a seamless form of global online education. In many cases the online courses can also help students in the home campuses by opening up new courses and professors to complement the on-campus programs.

5.5 Stanford Online Education

Stanford University has played a pivotal role in developing online learning since the 1960s when Stanford University began recording some of its engineering courses for nearby Silicon Valley companies. In the early days Stanford would deliver video cassettes of the courses to neighboring technical companies to enable corporate engineers to take courses remotely in something like a pre-historic "MOOC." Further advances in close-circuit television allowed live broadcasts of engineering courses on Stanford's campus to companies like Hewlett Packard. As Stanford alumni left campus, many played a key role in founding Silicon Valley's corporations, such as when Stanford graduate students Larry Page and Sergey Bin founded Google in 1998. The Stanford Computer Science department and Engineering school together have launched hundreds of students and faculty who founded startups in the area, including Andrew Ng, Daphne Koller of Coursera and Sebastian Thrun, who founded Udacity. Stanford Engineering Everywhere (SEE) was one of the first MOOC providers, founded by Andrew Ng before his work with Coursera. The SEE program still offers several computer science courses free of charge, including the famous Machine Learning course, CS 229 by Andrew Ng.

During our visit to Stanford's online learning group in 2013, we were greeted by nearly the entire online education team. In 2013, this included Robyn Dunbar (director of the Center of Teaching and Learning), Amy Collier (Director for Technology and Teaching for the Office of the Vice Provost for Online Learning), Jane Manning (Manager of Online Course Production and Platforms) and Brent Izutsu (Director of Digital Media), who all gave presentations. The Stanford online group seemed to have a real sense of excitement and energy, and their work was generously supported by the Stanford President and Provost. In 2012, Stanford's President, John Hennessy, described Stanford's new Online initiative as part of a tradition of leading online learning:

> Stanford has been at the forefront of this game-changing, challenging initiative. Our faculty have been working in online education for some time now, and their excitement is growing. This is a field that deserves increasing attention and investment, and the new Office of the Vice Provost is in keeping with Stanford's tradition of leadership in innovation and experimentation. (Stanford 2012).

The Stanford Provost, John Etchemendy, described how the online learning would both serve students on campus and advance the institutional mission to disseminate knowledge:

> Our primary mission is to teach Stanford students, but it is also the university's mission to disseminate knowledge widely, through textbooks, scholarly publications and so forth. Technology provides ways to both improve our existing classes and to extend our reach. By using technology creatively, we can tap the tremendous teaching talent we have on campus to offer new learning opportunities to millions of people, both in the United States and around the world. (Stanford 2012)

Stanford in March of 2013 was advancing into online education with a "full court press" that included simultaneously building two in-house online learning platforms (class2go and venturelab), running an online high school, supporting iTunesU courses, and also supporting Coursera. Stanford was also offering seed grants of $25 K to faculty for developing new online courses, and the encouragement from all levels means that a number of exciting new projects are being developed at once, even in fields not traditionally thought fit for online education like dance and humanities. The development of online education at Stanford was supported from the highest levels, and took a "many flowers bloom" approach. This approach was chosen to allow the best solution to emerge from a large number of competing systems in the early days.

Brent Izutsu, director of Stanford online education's digital media department, explained some ways in which production values can become important issues when developing content for an online course. Many online courses now offer more than a professor recording in his office, and include HD video, well-lit sets, and chromakey filming in studios. These high production values are now increasingly necessary to avoid complaints from increasingly sophisticated students. Robyn Dunbar noted that the entire process of online learning development and the phenomenon has energized discussions of teaching both inside and outside of Stanford, and "has brought new faculty to the table" that has helped many faculty want to work more at effective teaching. Online learning also has brought a new maturity to the process of course design. She marveled that for the first time, faculty are asking "who is my audience?" and thinking more deeply about how well their teaching is connecting with students.

Since 2013, Stanford has moved all of its online education initiatives within the office of the Vice Provost of Teaching and Learning, who is John Mitchell, himself a pioneer in computer science and online learning. Stanford continues to offer a wide range of online courses, and is one of the top three universities in the world for online courses. The online offerings include courses staged by Coursera, edX, Stanford edX, Udacity, and Stanford's own online learning platform, Lagunita. The diversity of platforms and offerings indicates that the "many flowers bloom" approach is still being pursued and is providing ample new data on how students learn and develop within online courses.

Educational research is of increasing importance for Stanford's online courses, and the VPTL houses several initiatives and centers for researchers who can both develop and study online learning. The Open Learning Initiative, for example, "is designed to simultaneously improve learning and facilitate research" and provides high-quality courses with advanced learning analytics. This allows for the courses to connect with "cutting-edge learning science" while enabling students to have improved feedback on their learning and help professors gain new insights into teaching and learning. The Center for Advanced Learning through Online Learning (CAROL) connects international partners and collaborators in research studies about online learning and has created a large and global community of scholars exploring ethical and effective practice within online learning. The Lytics lab is another initiative that explores the human factors within online learning, such as the ways that students can collaborate and form teams in online courses, as well as promoting gender equity and positive learning mindsets in online courses (Stanford VPTL 2017).

During a visit to Stanford's VPTL offices in 2018, I met with Marc Sanders, who was the Associate Vice Provost for Design and Innovation at the VPTL. Marc confirmed the multi-centered approach at Stanford, which still today takes the "many flowers bloom" approach. This includes work with the Center for Health Education to develop online medical courses and an online Master's program in epidemiology, and CAROL, which is doing a variety of research projects related to online learning. Marc stressed that Stanford is constantly improving its online education experiences, using both the Canvas LMS, and its various online learning platforms. Currently about 100 instructors are offering a variety of online courses, and they are using the new Lagunita platform, itself a variant of the open edX online learning platform. In all of their efforts, they are noticing that these online courses help students on campus by supplementing their on-campus experience, and also that the courses, by making a wide range of models of teaching visible, perform a very useful form of faculty development.

5.6 Udacity

Udacity was third of the major online course providers, and began within a short bike ride from Stanford's campus, as its founder, Sebastian Thrun, has deep Stanford roots. Thrun gained fame for his engineering genius in helping invent a self driving car, which won the 2005 DARPA Grand challenge, and which has been further refined by Google to drive in urban environments. Thrun was also a very popular teacher at Stanford, where his course on artificial intelligence enrolled over 160,000 students when it was offered online in 2011. The Udacity organization began in 2012, with seed money from venture capitalists providing $15 million in the first year. The first two courses were launched in early 2012, a course on building a search engine, and Thrun's course on "Programming a Robotic Car." Udacity's courses later branched outside of computer science to include courses across the STEM curriculum. By the end of 2012 more than a dozen courses were available in science, math and computer science, and in mid 2013, over 28 courses were available.

Udacity became famous and then somewhat infamous for its work with San Jose State, which promised to provide three online, for-credit courses for just $150 each. These courses were developed with the support of California Governor Jerry Brown. The courses were opened in an initiative entitled "SJSU Plus" with great fanfare during the Spring and Summer of 2013 (Straumsheim 2013), but on average less than half of the students passed the Udacity course, compared with 74% of the SJSU campus students. The pass rate for the three Udacity courses was 29, 44, and 51%. Ellen Junn, Provost of SJSU stopped the Udacity online for-credit course experiment, in response to these low pass rates (Harris 2013). A preliminary analysis suggested that the Udacity course was put together very quickly and included a number of students who were still in high school, who failed earlier math courses, and who had limited access to computers. Many of the students were also unaware of the online tutors, who could have helped prevent failure if they were better utilized (Corcoran 2013). Thrun was undaunted by the apparent failure, and claimed to be close to finding a "Magic Formula" to online education based on his experience. Thrun explained how the 'failure' of the SJSU experience was part of a natural cycle of innovation:

> To all those people who declared our experiment a failure, you have to understand how innovation works. Few ideas work on the first try. Iteration is key to innovation. We are seeing significant improvement in learning outcomes and student engagement. And we know from our data that there is much more to be done. (Straumsheim 2013)

We visited Udacity's headquarters during Spring of 2013, before the results from the SJSU experiment were available. The Udacity building from the outside looked very corporate but the interior had a whimsical and youthful atmosphere, with writing all over the walls and obvious signs of exuberant creativity. Whiteboard walls were covered with outlines of courses, cartoons and other creative ideas. A course outline on one wall outlined a genomics course with learning goals, and specific lectures broken down. Other cartoons were drawn on walls, and employee names were written with molecular type diagrams.

Our 2013 tour was led by Stuart Frye, who directs strategic partnerships and business development for Udacity. Stuart was a voluble 30-something, which made him a *senior* member of the Udacity staff. Stuart explained that the space was indeed "cooler than most startups" since this was their third office, and they learned from the past two. He explained how many of the employees loved coming to work and even stay late and on holidays because of the fun atmosphere. Stuart explained that Udacity took an approach to learning informed by the idea that the 2013 was similar to the time when film was first invented for motion pictures. In the beginning of film, movies primarily recorded theatrical performances before they recognized that new medium enabled productions entirely different from the earlier medium of live theatre.

The invention of the "set" and "location filming" both created what we identify as a modern movie, with entirely new processes for planning, production and post-production. Udacity brought many of these cinematic ideas to their course design, and aimed for a higher quality, cinematic type of course in the initial years (Fig. 5.4).

Fig. 5.4 Stuart Frye, director of strategic partnerships for Udacity, in 2013 explaining the "cinematic" nature of Udacity's online courses, at the company headquarters "set" for recording in-person shots

Each Udacity course was developed carefully with a team of production technicians. The instructor visited their headquarters and recorded in their studios using both a camera/set and a desk which recorded the instructor's writing as they narrated. Additional "location" segments made their way into some of Udacity's initial course offerings. Some of the courses also employed actors instead of just using professors who may be camera shy. Skits that dramatized physics or statistics sometimes used youthful Udacity employees and in some cases camera-shy profs only appeared in the background (on a TV in the living room) behind the onscreen, 20-something actors.

Other differences in the Udacity courses, compared to Coursera, was that they were "asynchronous," so that students could begin a course at any time and proceed at their own pace. Udacity also employed "adaptive learning" as well as automatic grading and feedback for the courses, which leveraged some of the artificial intelligence expertise of their founder, Sebastian Thrun.

During our 2013 visit, Udacity CEO Sebastian Thrun sat with us and explained some of the key differences between Udacity and other online courses, and demonstrated a physics course with imbedded simulations of a cannonball trajectory. He explained how Udacity was "student centered," which made the courses more responsive to the needs of a student. He described online learning as an antidote to many of the problems in academia, and predicted a "revolution" that will enable the professors to reach more students, and will improve teaching and make higher education more democratic, while giving universities a clearer sense of their mission.

Since 2013, Udacity has shifted its core mission away from free online university courses toward courses that link directly with employers and corporate training efforts. After the experiment with San Jose state, Udacity pivoted toward offering courses that would be directly relevant to students in their work, primarily in technical fields. Initial slogans for this new emphasis was "Be In Demand" and "Jobs of Tomorrow." In 2016 Udacity offered students an option known as "Nanodegree Plus" which guaranteed a job within six months, or the cost of the degree would be refunded. The four Nanodegrees with this guarantee were "Android Developer," "iOS Developer," "Machine Learning Engineer," and "Senior Web Developer," (Thrun 2016; Shah 2016). In 2017, Udacity hired a new CEO, Clarissa Shen, who in an interview further emphasized the shift away from MOOCs. She told the Economic Times in India that MOOs are "dead" and that "MOOCs are a failed product, at least for the goals we had set for ourselves. Our mission is to bring relevant education which advances people in careers and socio-economic activities, and MOOCs aren't the way" (Khosla 2017).

Udacity's emphasis has since shifted toward more international partnerships and a larger presence globally, as it established offices in over 10 different countries and hired hundreds of engineers to create country-specific web platforms, and to link more directly to companies and students in India, China and other countries. Career fairs and programs in India, Brazil, China and other countries further Udacity's emphasis on direct linkages between its online courses and employment. Udacity is also creating further nanodegrees to prepare students for the "Jobs of Tomorrow."

Among its recent new offerings are Robotics, Intro to Self-driving Car, Digital Marking, and even a Flying Car Nanodegree that includes former CEO Sebastian Thrun (who is also CEO of the flying car company Kittyhawk) and an MIT professor of Aeronautics and Astronautics. These nanodegrees are growing in popularity, and already have graduated tens of thousands of students (Shah 2017b).

5.7 EdX and Open-Source MOOCs

edX is the East Coast counterpart to the Bay area trio of Udacity, Coursera and Stanford. edX was founded in a partnership between Harvard and MIT and is led by Anant Agarwal, an MIT professor of Electrical Engineering and Computer Science. The first edX course, Circuits and Electronics, was offered by Agarwal in 2011, as part of MITx, and attracted 155,000 students in 162 countries. Harvard and MIT both contributed $30 million to launch edX in May 2012, and by 2013 the edX consortium grew to include 30 partner universities in 11 countries, with 86 courses that have been taken by over a million students, with a goal to eventually "educate a billion people." By 2017, edX had 14 million students, and offered 64 professional education credentials, and a series of 43 "MicroMasters" and 35 Professional Certificates, including an online Masters of Science in Analytics through Georgia Institute of Technology (Class Central 2017).

Before starting edX, Agarwal served as the director of MIT's Computer Science and Artificial Intelligence Laboratory (CSAIL). He also founded a company called Tilera that invented and manufactured new types of microprocessors. Agarwal received several awards for his achievements as a researcher in computer architecture and hardware. His educational background includes a PhD from Stanford and a Bachelor's degree from the Indian Institute of Technology, Madras. Agarwal's goal in founding edX was "to reinvent education" and he predicted that online education would be "the single biggest change in higher education since the printing press" (Rosen 2012). In several of his speeches he joked that a picture of MIT from 50 years ago has very little difference from now, except possibly the color of the seats. Agarwal relished the chance to use technology for education, which he said we should "embrace" since it is part of the "millennial generation's natural predilections." Agarwal hopes to integrate some of the latest educational research into his online courses—these elements would include active learning, self-pacing (so students can "rewind the professor"), instant feedback, gamification, and peer learning. The result will be a completely different form of technology-enhanced learning where the lecture hall of 50 years ago will be obsolete. Agarwal joked about this in a blog post:

> We will still need one lecture hall in our universities. Otherwise, how else do we tell our grandchildren that their grandparents sat in that room in neat rows like corn stalks? And didn't even have a rewind button. (Walters 2013)

On March 31, 2013, I visited the Kendall Square edX headquarters near MIT to meet with their team and talk with Anant Agarwal (Fig. 5.5). We met with Howard

Fig. 5.5 View of edX headquarters near MIT, showing a similarly youthful, and slightly denser and more intense environment than some of the California online companies

Lurie (former Vice President for Content Development) who described details of the edX organization. One question I asked was how edX was different from the other online learning companies. Agarwal pointed out that edX is non-profit and was designed to align with the principles of its partners—Harvard, MIT, etc.—all of which exist as non-profit institutions of *excellence*. These founding institutions were referred to as the "brand parents" of edX. Anant also stressed that edX is "guided by principles and not profit." This is embodied in their decision to offer the entire edX system to the world as an open source Python system.

Anant explained that edX was "very deliberate" about adding universities to their group. There are other unique aspects of the edX platform, which includes an advanced Artificial Intelligence Assessment engine, and the ability to arrange small group "cohorts" in classes, which is especially useful in humanities. Anant was also proud of "virtual laboratories" that were built into some of the edX courses and include both simulations and virtual laboratories. Some day these laboratories could integrate remotely operated telescopes, submarines and ROVs, and all manner of laboratory equipment that students could operate online.

Partners joining edX were asked to make a financial contribution to join, but they also could participate in revenue sharing, which varied by partner. The edX staff explained how developing online courses requires some training to help faculty deal with the fundamental paradigm shifts needed to implement an online course. The online course is not about filming someone in front of a class, but should reproduce the experience of being in a friendly office hour, with the professor talking directly to you and writing out things for you on a pad right in front of you. edX seeks out partner institutions that are able to achieve this kind of online course, by asking "hard questions" that are able to help professors "re-think the entire system they were educated with."

One interesting feature edX employed in its platform was to build a hierarchy of online students through "up-voting" of comments on online forums, and, at one point, edX discussion forums allowed learners to award "karma points" for especially helpful comments and answers in the discussion. One such student was a South African doctor who had been anonymously providing superb advice and comments in one of their courses. The quality of the online discussion added a new dimension to the courses, since the large volume of students across the world provides opportunities for both diverse views and expert comments.

Since 2013, edX has retained its position as a top provider of online education with strong partnerships with Harvard, MIT and other leading universities. Like Coursera and Udacity, edX has also shifted its emphasis toward stackable modular credentials (which edX called MicroMasters® program credentials) and online degree programs. By 2017, edX had developed over 170 credentials split across four different types that include MicroMasters programs, XSeries, Professional Certificates programs, and Professional Education, and is expanding its offerings—introducing over a dozen new online Master's programs in 2018. Like Coursera, edX is also offering corporate online learning called edX for Business, to enable companies to offer employees with a wide range of online courses and credentials. New for-credit courses are now being arranged for transfer credit at both US and international universities. An example is the Virtual Exchange program from TU Delft, where students from eight global universities can apply online edX courses from the other institutions to their own degree programs (Michelen 2017). Through the expansion of for-credit and paid options, edX is creating new markets and revenue sources, much like its rival Coursera.

In a speech at the 2018 Open edX conference at the University of Montreal, Anant Agarwal announced some ambitious goals for the next five years. Agarwal has the goal of adding 100 million learners, which is more than doubling the current number of online learners, and a huge increase from the 16 million learners on the platform during 2018. Agarwal also anticipates using more AI-enabled adaptive learning, which will allow the platform to adjust its lessons to each student individually. Agarwal also hopes Virtual Reality (VR) will become more common within the edX platform, and this already has happened in some training such as in medicine. With the incorporation of VR and AI, the online learning environment will offer experiences qualitatively different—and in some cases better—than is available within in-person and on-campus programs (Walker 2018).

As Agarwal put it, the future will "leverage fully the power of digital technology for learning." Agarwal also took note of the growth of the OpenedX platform, which by its open-source nature, has enabled an explosive growth of courses, learners and new capabilities. Currently the OpenedX platform has more learners than edX.org, and the huge community of programmers within OpenedX has enabled the platform to incorporate more innovations such as adapting for languages such as Arabic and Hebrew, and incorporating new advances of cognitive science into the platform, and improving the engagement with modifications for cohort-based learning. Agarwal has also divided the impacts of his platform into three categories—before, during and after campus—which he describes as "omnichannel" education. As the expansion of the online learning continues, Agarwal noted that for universities "an omnichannel approach will be critical." Examples of the omnichannel growth include Micro-Masters programs, and the Global Freshman Academy, a partnership with Arizona State University, which now enrolls over 440,000 students in 180 countries, and enables them to get a full year of ASU credit toward their degrees either at ASU or other universities (ASU 2019). Beyond the technology, online stackable, modular degrees and credits bypass the admissions process, which favors more wealthy students. Combined with new corporate programs that guarantee interviews for students who complete certain micro-masters programs, students anywhere can compete and thereby "break the shackles" of many of these entry mechanisms, creating a more democratic process.

5.8 HarvardX—Online Education Goes to Harvard

To get more perspective on how and why an elite institution develops courses for online education, I made a trip to Harvard's Bok Center for Teaching and Learning in 2013, which is also home of the HarvardX offices, the center for Harvard's online course development. Harvard and MIT together founded edX, and while edX is responsible for the platform development and technical side of the online courses, HarvardX was established for developing new online courses with Harvard faculty. At the Bok center, I met with Terry Aladjem, director of the Harvard Bok Center, along with John Girash, who is in charge of the STEM side of Harvard's teaching center, and Rob Lue, the Faculty Director of the Center.

Harvard began its online learning development with a major gift from the Hauser Foundation of $40 million for teaching and learning. This gift was used for new grants for faculty to develop courses through a "Hauser Grant" program (Walsh 2011). Rob Lue, Director of the Bok Center, described how the HarvardX was shifting Harvard's culture to become more aware of teaching. Rob said that the HarvardX project has had a "catalytic effect" that has "energized faculty broadly," resulting in a series of Town Hall meetings on online instruction where between 100 and 130 faculty discussed the details and techniques of teaching for hours. Rob was delighted to see the proliferation of conversations about teaching across all divisions and departments, making this online learning effort "a one Harvard thing." Rob also explained that Harvard joined the online learning effort for "scholastic opportunities" and noted that had to decide whether to "watch from the sidelines to see what wins" or instead be a player and lead some of the changes coming.

Like Stanford Online education, HarvardX is now housed in the Office of the Vice Provost for Advances in Learning, and is described as a "University-wide strategic initiative enabling faculty to create open, online courses for on campus and global learners and advancing research in the learning sciences" (Harvard 2016). Robert Lue, the Faculty Director of HarvardX, noted in a recent report that "HarvardX's faculty-driven approach continues to distinguish what we do among open online learning experiences," and the report lists the goals of HarvardX includes reaching global learners, enhancing on-campus residential teaching and learning and creating opportunities for learning sciences research (Harvard 2016). In 2018, HarvardX is continuing to add courses and students and now offers over 269 courses online, has 6.83 million registered users from 193 countries, of which 1.2 million are from U.N. Developing Nations. Like a lot of the students in online courses, a large number (320,569) are working professionals, or teachers (34%). The demographics within HarvardX also features young demographic, with 72% of HarvardX students being "millennials" (Harvard 2018).

In 2016, HarvardX and MITx released one of the most detailed studies of online education user demographics and usage available. The report analyzed 4.5 million participants in 290 courses for trends in course completion and other variables from 2012 to 2016 (Chuang and Ho 2016). As one of the critiques of online learning has been low completion rate, the report used certificate completion statistics among "participants" within 236 HarvardX and MITx courses without certification fees to remove the effects of cost. Among these 4.05 million participants, a median of 7.7% completed certificates, with wide ranges of completions between courses ranging from 0.2% in Field Theory to 34% in Chinese History. Even higher median completion rates of 30% were seen among students who initially intended to complete the course, and among students who had previously paid for a certificate, fully 60% of students complete a certificate. These completion rates are higher than earlier studies and the vast dataset allows researchers to further study which factors of demographics and course design have the largest effects on course completion.

The report also documents that the 4.5 million students are mostly male (66%), from outside of the United States (70%), and a large majority of them have already completed their bachelor's degree (73%), with a median age of 29. Examining the completion rates of the international students shows a large correlation between completion rate and country of origin, with students from countries with a low Human Development Index (HDI) achieving significantly lower completion rates than those from more wealthy countries. The study suggests that in these high HDI value countries "marketing, infrastructure, incentives, and supports are likely converging to enable participants to earn certificates." Exceptionally high completion rates are observed in Rwanda, however, which the report suggests could arise from "targeted outreach there from a joint partnership between edX and Facebook." The vast data within the report enables further analysis to determine how to improve completion rates and support for students in less wealthy countries to increase access to the courses and increase their global impact.

5.9 The Future of Online Learning and Colleges and Universities

In 2018, online learning has evolved substantially since the "year of the MOOC" and serves many of the needs of universities and students. Online courses can make a college or university more visible to potential students and donors, and help fulfill the mission of the institution to disseminate knowledge widely. In addition, online courses can provide broad access to high quality education for all types of students across the country, including those in under-resourced high schools. The combination of online education with in-person meetups or hybrid online and in-person classes led by faculty can help improve outcomes for diverse student audiences with varying levels of preparation and motivation. By leveraging online technologies, higher education can help solve some of the perennial problems with cost and be more responsive to the needs of our society.

As we enhance diversity within our on-campus student populations, online education can expand our cultural and geographic diversity by bringing additional voices and perspectives into our courses. Online learning providers are increasingly involved in fostering global partnerships between universities, which creates global communities of students, faculty and companies. As this community becomes more global, international providers of courses are beginning to compete with the major US providers, and can offer new ways of reaching billions of students. In 2018, the top five providers of online learning include the three companies we visited in 2013 (Coursera, edX and Udacity—ranked #1, 2, and 4), but also two newer international companies as well—XuetangX from China, which has 9.3 million users, and FutureLearn from Europe, with 7.1-million users. The global online learning companies are also joined by India's new online provider, SWAYAM, which launched in 2016, and soon will reach all the universities of India with on-credit courses, potentially offering tens of millions of new students in the world's most populous nation.

By creating online courses and sharing the best forms of teaching at our leading universities, it is possible to stimulate discussions about teaching both within and beyond the campus. In many cases, online courses showcase the best instructors on the best campuses in the world, giving a much wider audience among both teachers and students. The unique demands of developing an online course have also pushed some of the curricular boundaries to explore topics outside of the traditional curriculum, and to break free some of the limitations of lecture-format classes with built-in interactive exercises, simulations and exciting on-location footage.

Designing new forms of online education that not only engages students as consumers of content, but also as producers and co-creators of new course units and group projects has the potential to provide rich learning environments for online learning that can extend the campus community across the earth to enrich discussions with extremely diverse cultural perspectives. With the accumulation of experience—now over 5 years with millions of students—the creators of online courses can improve how they approach their subjects and increase the persistence and completion rates within students. Online courses are reaching an international population that is more diverse and more educated that our on-campus undergraduate students, which enables online education to be optimized to reach students outside of campus after graduation. This can serve a strong vehicle for improving alumni relations, and also can extend the mission of a college to provide lifelong education. As Nikhil Sinha, former Chief Content Officer at Coursera put it, "College education used to be a 'step function' and now it is more like a 'wave'—as students can come and go over a longer period."

The online courses can help campuses reach their students throughout their lives and help build this more 'wavelike' experience of higher education.

The evolution of the existing platforms will enable dramatic new capabilities that will better emulate the experience in small classes and even laboratory sessions across time and space in both synchronous and asynchronous courses. For now, courses strive to be "engaging" but only the campus experience is able to provide hand-on experiences that are not possible online. Few would have predicted the role that web sites and the internet play in daily life in higher education 25 years ago, when web sites were only found at exotic particle physics and NASA laboratories. Web sites now have matured to a point where a College could not do any of its business without the web, as everything from admissions to enrollment management to library collections are online, and most courses have LMS web sites with blogs, streaming video and other features that would have seemed outlandish 20 years ago!

This observation leads me to a pretty certain prediction—whatever online learning looks like today is only the beginning, and in 25 years our College and University system will be similarly interwoven with online courseware to a point where it would be difficult to imagine teaching without some of the currently exotic features as online quizzes, imbedded video lectures, interactive simulations and instruments, and advanced threaded discussion forums that are all part of the online platforms.

References

ASU. (2019). *Global freshman academy*. Retrieved January 14, 2019, from https://gfa.asu.edu/.

Chuang, I., & Ho, A. (2016). HarvardX and MITx: Four years of open online courses–fall 2012-summer 2016. *Available at SSRN 2889436*.

Class Central. (2017). *By the numbers: MOOCs in 2017*. Retrieved January 14, 2019, from https://www.class-central.com/report/mooc-stats-2017.

Corcoran, B. (2013). Udacity's lessons learned. *EdSurge Online Journal*. Retrieved January 14, 2019, from https://www.edsurge.com/n/2013-07-24-udacity-s-lessons-learned.

Downes, S. (2008). Places to go: Connectivism & connective knowledge. *Innovate: Journal of Online Education, 5*(1).

Farr, C. (2013). Coursera nabs $43 M to bring online education to emerging markets. *Venturebeat Online Journal*. Retrieved January 14, 2019, from http://venturebeat.com/2013/07/10/coursera-nabs-43m-to-bring-online-education-to-emerging-markets/.

Harris, P. L. (2013). SJSU plus: Fall 2013 update. *SJSU Today*. Retrieved January 14, 2019, from http://blogs.sjsu.edu/today/2013/sjsu-plus-fall-2013-update-2/.

Harvard. (2016). *HarvardX: Year in review 2015–16*. Retrieved January 14, 2019, from http://harvardx.harvard.edu/files/harvardx/files/harvardx_15-16_year_in_review_-_10-24-16_0.pdf.

Hill, P., & Feldstein, M. (2013, April 3–4). Everything you thought you knew about MOOCs could be wrong. *ELI Online Spring Focus Session*. Retrieved January 14, 2019, from https://events.educause.edu/eli/focus-sessions/2013/eli-2013-online-spring-focus-session/proceedings/everything-you-think-you-know-about-moocs-could-be-wrong.

Khosla, V. (2017, October 6). Udacity to focus on individual student projects. *The Economic Times* (India). https://economictimes.indiatimes.com/industry/services/education/udacity-to-focus-on-individual-student-projects/articleshow/60963078.cms.

Koller, D. (2012). What we are learning from online education? *TED Talk*. Retrieved January 14, 2019, from http://www.ted.com/talks/daphne_koller_what_we_re_learning_from_online_education.html#201598.

Kolowich, S. (2013, February 7). American council on education recommends 5 MOOCs for credit. *The Chronicle of Higher Education*. Retrieved January 14, 2019, from http://chronicle.com/article/American-Council-on-Education/137155.

Lewin, T. (2013). California bill seeks campus credit for online study. *New York Times*. Retrieved January 14, 2019, from http://www.nytimes.com/2013/03/13/education/california-bill-would-force-colleges-to-honor-online-classes.html.

Michelen, A. (2017). High-ranking universities to sign a virtual exchange agreement. In *Engineering 360 from IEEE GlobalSpec*. Retrieved January 14, 2019, from https://insights.globalspec.com/article/7313/high-ranking-univerisities-to-sign-a-virtual-exchange-agreement.

Milligan, C., Littlejohn, A., & Margaryan, A. (2013). Patterns of engagement in Connectivist MOOCs. *MERLOT Journal of Online Learning and Teaching, 9*(2).

Rosen, R. (2012, May). The single biggest change in education since the printing press. *The Atlantic*.

Shah, D. (2015). *Eyeing revenue sustainability: The two biggest MOOC providers adapt how their courses work*. Retrieved January 14, 2019, from https://www.class-central.com/report/coursera-paywall-edX-discontinues-free-certificates/.

Shah, D. (2016). *Udacity's nanodegree plus comes with a job guarantee*. Retrieved January 14, 2019, from https://www.class-central.com/report/udacitys-nanodegree-plus-comes-job-guarantee/.

Shah, D. (2017a). *A product at every price: A review of MOOC stats and trends in 2017*. Retrieved January 14, 2019, from https://www.class-central.com/report/moocs-stats-and-trends-2017/.

Shah, D. (2017b). *Udacity shows strong growth, launches flying car nanodegree with MIT faculty*. Retrieved January 14, 2019, from https://www.class-central.com/report/udacity-flying-car-nanodegrees/.

Siemens, G. (2005). Connectivism: A learning theory for the digital age. *International Journal of Instructional Technology and Distance Learning, 2*(1), 3–10. Retrieved from http://www.itdl.org/Journal/Jan_05/article01.htm.

Sinha, N. (2018). 6 new degrees coming to coursera in computer science, data science, and public health. *Coursera Blog*. https://blog.coursera.org/6-new-degrees-coming-to-coursera-in-computer-science-data-science-and-public-health/.

Stanford. (2012, August 30). Stanford takes landmark step in online learning, appoints new vice provost. *Stanford News*. Retrieved January 14, 2019, from http://news.stanford.edu/news/2012/august/online-learning-office-083012.html.

Stanford VPTL. (2017). *Data-driven research—Accelerating and scaling the science of learning at CAROL and the Stanford Lytics lab*. Retrieved January 14, 2019, from https://vptl.stanford.edu/news/data-driven-research-accelerating-and-scaling-science-learning-carol-and-stanford-lytics-lab.

Straumsheim, C. (2013 August 28). Boost for udacity project. *Inside Higher Education*. Retrieved January 14, 2019, from http://www.insidehighered.com/news/2013/08/28/san-jose-state-u-posts-improved-online-course-results-udacity-partnership-remains.

Thrun, S. (2016). *A new job guarantee*. Retrieved January 14, 2019, from https://blog.udacity.com/2016/01/a-new-job-guarantee.html.

Walker, J. M. (2018). Relive the best moments from open edX2018. *Open edX Blog*. Retrieved January 14, 2019, from https://open.edX.org/blog/relive-best-moments-open-edX-2018.

Walsh, C. (2011, October). Harvard receives $40 million for innovative teaching and learning. *Harvard Gazette*. Retrieved January 14, 2019, from http://news.harvard.edu/gazette/story/2011/10/education-and-innovation/.

Walters, H. (2013). Reinventing education for millenials: Anant Agarwal at TEDGlobal 2013. *TED Blog*. Retrieved January 14, 2019, from http://blog.ted.com/2013/06/14/reinventing-education-for-millennials-anant-agarwal-at-tedglobal-2013/.

Wiener, J. (2013, September). Inside the Coursera Hype Machine. *The Nation*. Retrieved January 14, 2019, from http://www.thenation.com/article/176036/inside-coursera-hype-machine.

Chapter 6
Interdisciplinary Science

Abstract Numerous distinguished organizations, including the AAMC, AAAS, NSF, and NRC, have highlighted interdisciplinary research and education as vital for training future scientists and physicians. These recommendations are described in detail along with some examples of why interdisciplinary STEM education is needed for the 21st century. A review of a broad range of theories of multidisciplinary, interdisciplinary, transdisciplinary, and integrated science education is described, along with a taxonomy of methods for integrating curriculum in interdisciplinary or multidisciplinary programs. A review of interdisciplinary science programs ranging from one semester to four-year durations gives examples of this new kind of education can be implemented, and how interdisciplinary science has transformed the landscape of science education. Case studies from a set of 12 interdisciplinary programs in 5 countries provide detailed information on the curriculum and the ways in which faculty prepare for teaching these programs. From the review of these 12 programs a set of "best practices" for interdisciplinary science is presented, with detailed recommendations for structuring faculty training and for organizing the curriculum and assessment of teaching and learning.

Keywords Interdisciplinary science · Transdisciplinary science · Faculty development · Curriculum design

6.1 Integrated Science, Multi and Interdisciplinary Curriculum

In recent years, the most pressing and urgent issues of science and technology are inseparably complex problems that draw from multiple disciplines. Development of new sources of energy, studying the effects of global warming and sea level rise, and the effects of human population increase on rapidly evolving microbial communities and decreasing biodiversity in the natural environment are all interdisciplinary problems that demand solutions. Training students to effectively contribute to solutions in these areas, either as scientific professionals, or as informed policymakers and citizens, requires a retooling of our science education beyond the single subject elective course or disciplinary curriculum.

B. E. Penprase, *STEM Education for the 21st Century*,
https://doi.org/10.1007/978-3-030-41633-1_6

The urgency of interdisciplinary science education for training the next generation of STEM practitioners has been recognized by a number of societies and academies. The American Cancer Society report entitled *The Role of the Private Sector in Training the Next Generation of Biomedical Scientists* states:

> In the postgenomic era of research, multidisciplinary and interdisciplinary research will command center stage, requiring team approaches and the collaboration of many individuals from vastly different fields, ranging from computational mathematics to clinical science. (American Cancer Society 2000)

The National Research Council also noted that the changing needs for biomedical and behavioral scientists required new approaches in STEM education and recommended that the "NIH should expand its emphasis on multidisciplinary training in the biomedical sciences" (National Research Council 2000). The Bio2010 report, *Transforming Undergraduate Education for Future Biologists*, suggests that new approaches for higher education should consider "building a strong foundation in mathematics, physical and information sciences to prepare students for research that is increasingly interdisciplinary in character" (NRC 2003). A AAMC report (2004) on Medical Education in the United States highlighted the importance of interdisciplinary education for future medical practitioners, and noted these professionals need to be ensured to have the "knowledge, skills, attitudes, and values needed for medical practice as members of an interdisciplinary health care team, and the ability to perform the complex, integrative tasks required to provide high-quality health care to patients who seek their help" (AAMC 2004). The AAMC-HHMI report *Scientific Foundations for Future Physicians* strongly endorsed a movement toward more interdisciplinary STEM education, stating that "it is the committee's belief that at both levels—medical school and undergraduate schools—interdisciplinary approaches are an important component of the needed new directions in science education." (AAMC 2009). The AAAS report *Vision and Change in Undergraduate Biology Education* noted how interdisciplinary fields have produced many of the most exciting discoveries in biological sciences. The report states:

> "Emerging interdisciplinary fields such as genomics, proteomics, metagenomics, synthetic biology, biochemistry, bioinformatics, computational biology, and systems biology are leading to new discoveries, and some are changing the ways we think about and engage in biological research and explore established biological fields (such as evolutionary biology). These new integrated fields, spread across the diversity of life sciences, are opening up a vast array of practical applications, ranging from new medical approaches, to alternative sources of energy, to new theoretical bases in the behavioral and social sciences." (AAAS 2011)

Despite the strong support for interdisciplinary work, including from the National Science Foundation which supports "bold interdisciplinary projects in all NSF-supported areas of science and engineering research," science faculty and their courses are generally confined to single disciplines, which stifles interdisciplinary research and teaching. In order to conduct this work, the NSF notes that "a cultural shift in the mind-set is required to promote interdisciplinarity both within the research ecosystem and within academia" (Global Research Council 2016).

6.2 The Meaning of Interdisciplinary Science

Science teaching has traditionally been confined into "silos" where instruction adopts the methodologies and terminologies of each of the disciplines and generally focuses on single subjects in isolation from other STEM disciplines. This approach results in divergent and linear curricular paths, where each of the STEM subjects run in parallel, with little overlap or cross-connections. A slight improvement on this situation is what might be called a "multi-disciplinary" approach, in which each of the STEM disciplines are taught, but instructors build into the curriculum intentional cross connections between disciplines, allowing for the usual parallel and silo-ed instructional paths to intersect on a key case study or application (Fig. 6.1). Interdisciplinary education, as opposed to multi-disciplinary education, includes instructional paths that are interwoven around common topics or themes that can be studied with multiple disciplines (Fig. 6.2), building deep understanding of the connections within and across disciplines (Bloom 2004).

An interdisciplinary approach will develop the processes, skills, and concepts from two or more disciplines at the same time, using common themes or modes of inquiry to form interdisciplinary connections. An integrated or integrative approach would bring those same interdisciplinary elements together within a curriculum, but include a strong emphasis on inquiry-oriented pedagogy, and raise the theme or topic considered as the primary element within the curriculum, to which the disciplines are secondary, and in which the exact role of each discipline is "lost" to the search for answers to larger questions that are intrinsically non-discipline specific (Mathison and Freeman 1998).

The interdisciplinary approach may be likened "to a seamless woven garment that stands in contrast to the patchwork quilt of multidisciplinary work." In such an environment, "a concatenation of disciplinary perspectives is replaced by integration of those perspectives" (Lattuca 2001). Some have also made the distinction

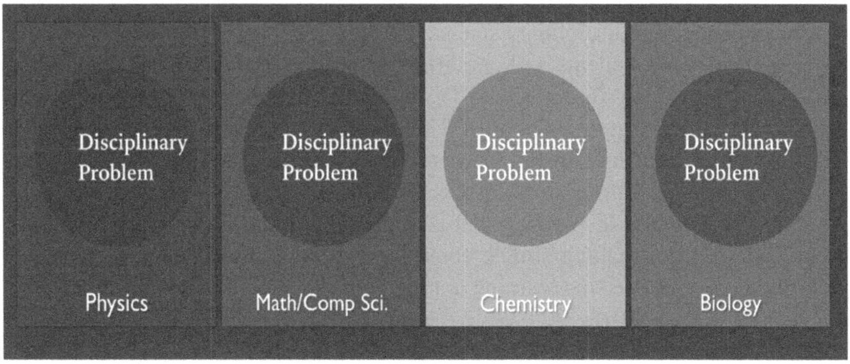

Fig. 6.1 Schematic of a multi-disciplinary curriculum in which a series of disciplines are sequentially engaged to solve disciplinary problems and to give students exposure to multiple disciplines

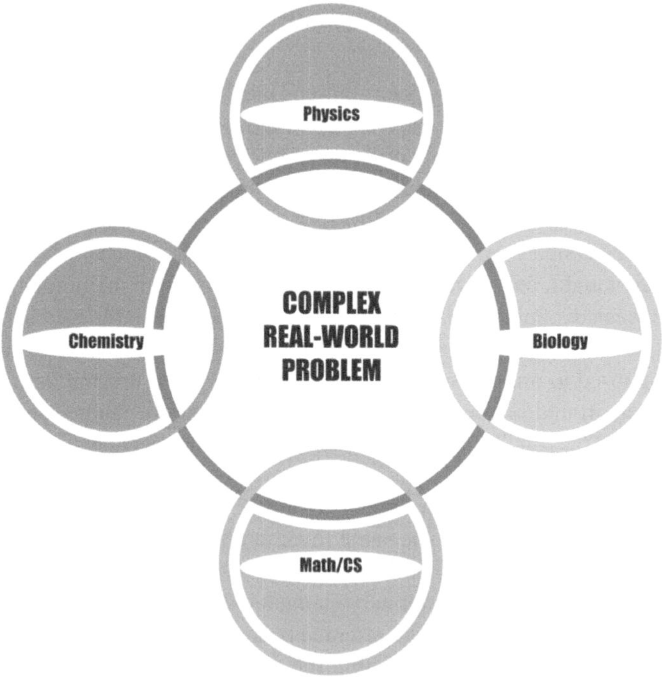

Fig. 6.2 Schematic of an interdisciplinary curriculum which enables multiple disciplines to converge on a complex problem simultaneously using multiple modes of inquiry to solve the problem

between instrumental interdisciplinarity, in which the interdisciplinary techniques are pragmatic problem-solving approaches, or conceptual interdisciplinarity, which emphasizes the synthesis of knowledge, which becomes an "epistemological enterprise involving internal coherence, the development of new conceptual categories, methodological unification and long-term research and exploration" (Lattuca 2001).

A coordinated interdisciplinary science program can provide for deeper mastery and engagement for students and can bring within the students a valuable awareness of how previously disconnected facets of information can apply to complex problems through integrative learning. The development of this type of integrative learning within a curriculum has been recommended by the American Association of Colleges and Universities (AAC&U), who describe this form of learning as "an understanding and disposition that a student builds across the curriculum and co-curriculum, from making simple connections among ideas and experiences to synthesizing and transferring learning to new, complex situations within and beyond campus" (AAC&U 2010).

6.3 Curricular Integration Within Interdisciplinary Programs

Without careful management, an extended STEM curriculum across a multi-year program with multiple instructors and disciplines has been likened to a "jigsaw puzzle," which for students "the curriculum is a pile of jigsaw puzzle pieces without a picture" (Beane 1995). Curriculum integration can enable linkages between disciplines and thereby allow for higher order learning. Examples of successful integration within professional degree programs in pharmacy (Pearson and Hubball 2012) and in medical education (Harden 2000) provide useful principles that can guide planning for multi-year interdisciplinary STEM curriculum. The sub-field of Scholarship of Curriculum Practice (Hubball and Gold 2007) can help inform the integration of teaching and learning, course design, faculty development and institutional strategy.

Several authors have developed frameworks for integration that consider the learning context, and the interrelations between planning, assessment and programming (Hubball and Burt 2015). Taxonomies of curriculum integration have been developed to describe how a curriculum progresses in phases from "fragmented" to "threaded," "integrated," "immersed," and "networked" integration (Fogarty 1991). The "integration ladder" can be used to visualizes stages of integration from isolated disciplinary perspectives toward transdisciplinary approaches (Harden 2000). Progress toward integration requires instructors to be aware of what is discussed in other courses; harmonization, where teachers responsible for different courses collaborate and discuss frequently to enable multiple connections to arise. The integration can then proceed from a "sequenced" set of courses toward "nesting", in which instructors draws from skills developed in other courses in their subject, or toward "temporal coordination," in which separate disciplinary topics are taught in parallel and students can uncover the relationships between the topics. Figure 6.3 shows a schematic of these initial curriculum integration levels, based on the taxonomies described in Harden (2000) and Fogarty (1991).

More intensive curriculum integration models include correlation, where disciplinary topics are discussed but time is set aside weekly or at the end of a semester for an integrative case study; complementary programs, where a theme connects multiple disciplinary courses and disciplinary experts to merge in solving an interdisciplinary problem. Integrated courses can provide approaches that range from multi-disciplinary program to interdisciplinary or even transdisciplinary, depending on the degree of emphasis on the connections between disciplines and their applications toward real-world examples that defy disciplinary classification (Harden 2000). Figure 6.4 provides a schematic of these more intensive forms of curriculum integration.

An integrated curriculum can benefit from curriculum mapping (Oliver et al. 2007), and should also be aware of the need for both horizontal and vertical integration (Malik and Malik 2011). Particular techniques include the use of working groups or faculty learning communities to divide work and ensure involvement from a wide

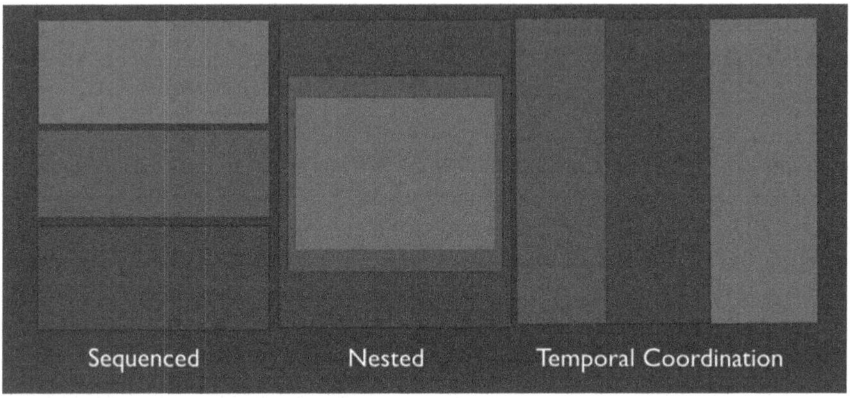

Fig. 6.3 Schematic of curriculum integration taxonomies where time in a course would move in the vertical direction. Disciplines (colors) can be in a sequence (left), a "nested" configuration (center) or taught in parallel (right)

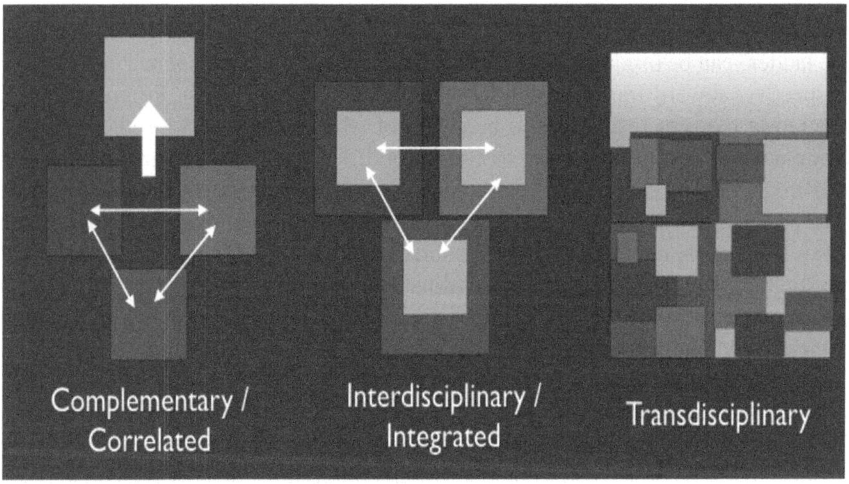

Fig. 6.4 Schematic of curriculum integration where multiple disciplines are connected through a common focus such as a case study (left), integrated within a program (center), or achieve transdisciplinary integration (right) (*Source* Harden 2000 and Fogarty 1991)

range of stakeholders to develop learning outcomes and skills, to provide meaningful links between disciplines, and to set a timetable to assure completion (Hubball and Burt 2015). Several authors stress the need for strategies and timetables for assessment, communication, and re-evaluation and revision, as well as for providing sufficient resources (including incentives for faculty) and a method for documenting improvements (Pearson and Huball 2012; Richlin and Essington 2004). Networked Improvement Communities can guide and visualize the curriculum development and integration, using an "improvement map" to limit the tendency to offer multiple and competing interventions, and offer an "end to end description of the challenge space" with "driver diagrams" linking the targeted improvements with causes for underperformance, and with specific assessable solutions (Bryk et al. 2010).

6.4 Overview of Global Interdisciplinary Science Curricula

Over the past decade, several institutions have responded to the need for interdisciplinary training with new curricula designed to foster interdisciplinary science practitioners. The Claremont Colleges W.M. Keck Science Department offered its Accelerated Integrated Science Sequence (AISS) from 2007 to 2017, funded initially by an NSF STEP grant to prepare students interested in medicine or science more efficiently in their introductory courses. The AISS course combined three one-year courses in physics, chemistry and biology into a single double-credit course. The three instructors within the course provide a form of Just-in-Time Teaching (Simkins and Maier 2009), by sharing duties within the classroom and trading off each other's expertise in answering questions and describing interconnections between the disciplines of physics, chemistry and biology. Further details of the AISS program can be found in Purvis-Roberts et al. (2009), and Copp et al. (2012). Princeton University has developed the venerable Integrated Science program, where a small and select group of 20–30 students experience a three-year integrated sequence that includes over 20 instructors who teach high-level physics, math, computer science, molecular biology, and genomics. The Princeton program is designed as an elite cadre of future researchers, and has one of the most intense programs available. In recent years however, the scope of the Princeton program has been reduced, and now has been recast as an intense first year experience that fulfils the first-year requirements for chemistry, physics, computer science, and molecular biology (Princeton University 2017).

In addition to programs for future science researchers, a number of single year or multiyear science curriculum for future engineers or non-science majors have been developed. These programs include the Virginia Tech Integrated Science curriculum, which includes a two-year double-effort course sequence to give fundamentals of biology, chemistry, physics and math. The course is a gateway to Science majors and is intended to provide an accelerated and enhanced "active-learning environment stressing multidisciplinary modes of thought" that can enhance science education but also make connections between science, law, engineering and art (Virginia Tech 2017). Northwestern University has an Integrated Science Program, which offers accelerated and smaller disciplinary courses and a dedicated Integrated Science Program house where students can interact with each other and with program staff (Northwestern 2017). James Madison University offers an ABET accredited Integrated Science and Technology (ISAT) program in which science, mathematics, technology, and management and manufacturing engineering prepare students for careers that blend science, technology and business. The ISAT program includes courses that provide the social context of technology and sciences, as well as a four-course Holistic Problem-Solving Spine that allows students to apply system thinking to complex problems (James Madison University 2017). At the University of Massachusetts, Amherst, a program known as iCons (Integrated Concentration in Science) provides a selective integrative undergraduate science program within the university that focuses on two theme areas—Renewable Energy and Biomedicine/Biosystems. The program extends over the full four years of the undergraduate program and supplements the existing STEM majors with "opportunities to engage with real world issues as part of interdisciplinary teams" (UMass Amherst 2017a, b).

Throughout Asia and the Middle East several entirely new universities are being created, while in Hong Kong the undergraduate degree programs were recently expanded from three to four-year durations. These changes have created many new innovative interdisciplinary science courses and course sequences. Within the Core Curriculum at NYU Shanghai is the Foundations of Scientific Inquiry program, which includes a three-year sequence of disciplinary courses that starts with Quantitative reasoning, and progresses through Core science courses in Physical Science and Life Sciences that are designed to provide broad perspective on current topics and concepts in both subjects. The Physical Science courses discuss broad topics such matter and energy, climate, particle physics and dark energy, while Natural Science courses explore the intersection of neuroscience, information science, and genetics in the life sciences (NYU Shanghai 2017). Chinese University of Hong Kong has designed a single semester course entitled "In Dialog with Nature" that explores the methods of inquiry in science and mathematics from Euclid and ancient Chinese science to readings from Charles Darwin, Rachel Carson, and James Watson (Chinese University of Hong Kong 2017). University of Hong Kong offers a Fundamentals of Modern Science course, which is designed to give an "an overview of the giant web of knowledge that makes up science." The course integrates physics, astronomy, earth sciences, chemistry and biology in two semesters, with ambitious units that outline the "universal principles and unifying concepts of science", and apply them to the fundamental structure of matter, atoms, DNA, cells, organisms, earth, solar system and cosmology (University of Hong Kong 2017).

Yale-NUS College in Singapore was founded by Yale University and the National University of Singapore with a core curriculum that was designed to answer the question "what must a young person learn in order to lead a responsible life in this century?" (Yale-NUS College 2013). The answer is an ambitious core curriculum that includes courses in Quantitative Reasoning, Scientific Inquiry, and additional interdisciplinary science courses taken by all of its students. In its initial years, the Yale-NUS College STEM curriculum featured an intensive interdisciplinary science course known as Integrated Science taken by science majors, and a year-long interdisciplinary science sequence entitled Foundations of Science. Both courses were taught by teams of faculty and all Yale-NUS students would take at least four full semesters of STEM courses, which were geared toward interdisciplinary problems related to the theme of "water" in Integrated Science and "Grand Challenges" of civilization in Foundations of Science.

Other interesting programs within Canada and Britain include the Honours Integrated Science program at McMaster University, an interdisciplinary, research-based science program targeted toward future researchers. The program includes levels that progress in each year toward more complex research questions, starting with supervised inquiry-based learning, based on projects, followed by thematic modules emphasizing overlapping content between disciplines. The final stage is a fourth year Honours Thesis project (McMaster University 2017). The University of Leicester offers the Centre for Interdisciplinary Science, which includes innovative courses that incorporate research and culminate in a Natural Sciences degree (University of Leichester 2017). The University of British Columbia offers a program known as Science One, which is an integrated first-year science experience taught by 8 instructors, that offers 75 students integrated courses in biology, math, physics and chemistry, including a Marine Biology research experience at their own Bamfield Marine Sciences Centre (The University of British Columbia 2017).

6.5 Learning More—Interviews with Interdisciplinary Science Leaders

To learn more details about the dazzling array of global interdisciplinary science programs, a set of thirteen interviews were conducted with leaders of twelve interdisciplinary science curriculum efforts in five countries to determine how they have responded to the common challenges in organizing scientists with various disciplinary skills and vocabularies into a unified teaching team (Fig. 6.5) (Table 6.1).

A standard set of questions were asked of each of the leaders of the 12 programs, to learn more about the details of organizing and executing the interdisciplinary teaching and curriculum. These details are important, since faculty are trained in disciplines, and creating effective teaching teams that can work across their disciplines, and master the terminologies and methodologies of the other disciplines is difficult work. The questions were posed to learn more about how faculty were prepared for this task, and how they collaborated during the term to align their practices in grading and share knowledge about their disciplines and their teaching within the teaching team. The questions are listed in Table 6.2.

6.6 Case Studies of STEM Interdisciplinarity

The interviews not only revealed many fascinating details about the curriculum, but also about the importance of "human factors" within science teaching, and the need to knit together a team that feels a sense of common purpose in their work. In many cases, the interviews revealed some "best practices" that can be very helpful for other instructors who are designing or working within interdisciplinary teams. These insights are summarized in short case studies for many of the programs.

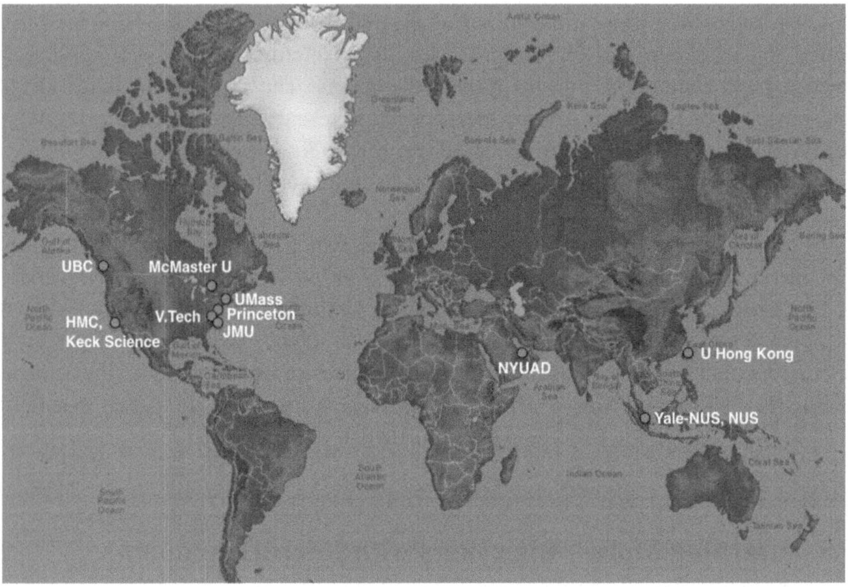

Fig. 6.5 Global distribution of interdisciplinary science programs considered in this study

Table 6.1 Interdisciplinary science professors contacted for this study

Interviewee/Reporter	Institution	Country	Course/Program
Joshua Shaevitz	Princeton	USA	Integrated Science
Scot Gould	Keck Science	USA	AISS
Katie Purvis Roberts	Keck Science	USA	AISS
Eric Maslen	JMU	USA	Integrated Science and Technology
Scott Auerbach	University of Mass	USA	iCONS
Darryl Yong	Harvey Mudd	USA	Probing Inverted Classroom
Michel Pleimling	Virginia Tech	USA	Integrated Science
Christopher Addison	University of British Columbia	Canada	Science One
David Scicchitano	NYU Abu Dhabi	UAE	Foundations of Science
Jan Gruber	Yale-NUS College	Singapore	Integrated Science
Shaffique Adam	Yale-NUS College	Singapore	Integrated Science
Adrian Lee	NUS	Singapore	Special Programme in Science
Bryan Penprase	Yale-NUS College	Singapore	Foundations of Science
JCS Pun	University of Hong Kong	Hong Kong	Fundamentals of Modern Science
Carolyn Eyles	McMaster University	Canada	iSci

Table 6.2 Interview questions for interdisciplinary science study

• How do you prepare faculty to teach in a multi-disciplinary environment?
• How do you share information on what each instructor is doing in their section?
• What techniques work best for professors to share pedagogical approaches?
• How can you assure consistency of grading within each section?
• How can you assess the quality of teaching and learning within the section?

6.7 Keck Science Department Claremont Colleges—AISS

The Claremont Colleges W.M. Keck Science Department offered an Accelerated Integrated Science Sequence (AISS) to prepare pre-medical and science majors for physics, biology and chemistry in an integrated year-long double course sequence. Funded by an NSF STEP grant, and now the Keck Foundation, the AISS program was offered to students from Claremont McKenna, Pitzer, and Scripps Colleges. The course was team-taught by three professors, representing the disciplines of physics, chemistry and biology, who offer an intense immersion into all the three sciences to their small class of 25–32 students. The course had 2 h meetings 5 days a week, and a separate 4 h lab—giving a total of 12–14 h per week of class time. A typical class was taught "studio mode"—with in-class lab exercises mixed seamlessly with lecture and discussion.

The AISS course included topics such as waves (which leads to discussion of the ear, electromagnetic radiation, and quantum mechanics), molecular structure (which leads to discussion of the physics of electrons, chemical bonding, and the shape of macromolecules) and energy within thermodynamics, organisms and cells. The topics within the AISS course were refined over the 10 years to include those that work well in the team-taught "studio" mode, and that naturally integrate the different science disciplines (Fig. 6.6) (Purvis-Roberts et al. 2009).

In AISS, all three instructors taught in the class at the same time, and frequently hand off discussion of phenomena to each other to allow a discussion of how each of the various disciplines (physics, chemistry and biology) apply to the topic. In one class, for example, students used the bioinformatics.org database and the RCSB Protein Data Bank on laptops to study biological macromolecules. The biology professor led the class, but the physics professor often would jump in and explain how electron polarization contributed to hydrophobic and hydrophillic parts of macromolecules. Additional information from the chemistry professor clarified how these macromolecules are formed and found in cells. Throughout the class students would break into teams to find and render large molecules on their computers using the databases and a three-dimensional visualization program in MAPLE.

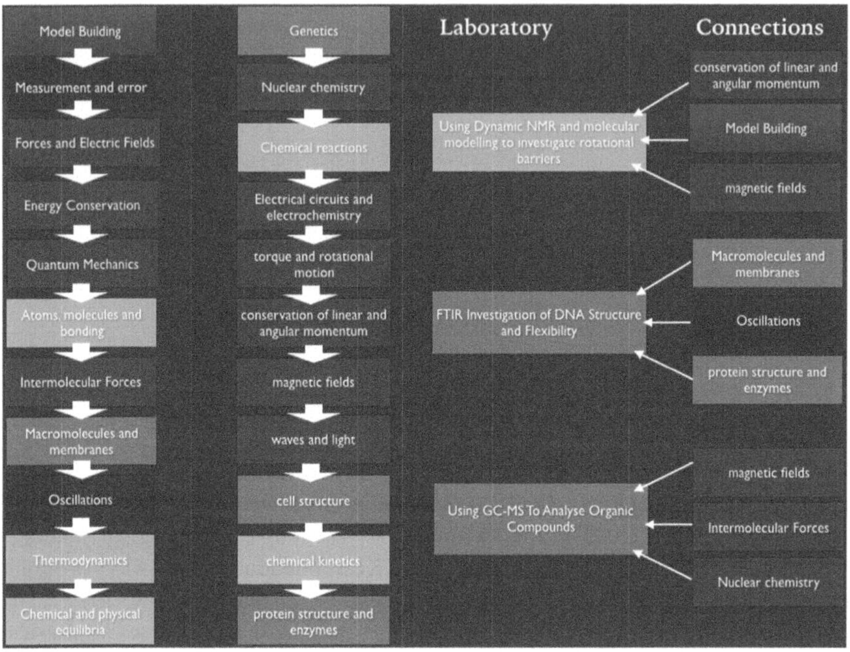

Fig. 6.6 Flow chart for topics in the Claremont Accelerated Integrated Science Sequence (AISS) classroom (left columns), and a sampling of lab topics in the course with connections to topics in class (right). Disciplinary topics include physics (red), biology (green), and chemistry (blue). *Source* Purvis-Roberts et al. (2009)

One key point that came out in the interview was how teaching in a course like AISS requires an open-minded approach. Each component of the course, laboratory schedule, topics to cover, and assessments have to be questioned and the purpose of each identified. This prevents instructors from repeating past practice without the deeper rationale for why each element exists within a class. The team-taught dynamic was also valuable for stimulating learning for both the faculty and students. With three teachers in the room often one of the instructors can raise their hands to provide an alternative viewpoint, and to also share a different way to teach a topic. Our interviewee, Scot Gould, shared that AISS was "the single most exciting thing he has done pedagogically, ever." He is convinced that this program is worthwhile for students entering medical school, as the AISS instruction is much more like the medical school environment which is typically interdisciplinary.

6.8 University of Massachusetts, Amherst—iCons Program

The University of Massachusetts Integrated Concentration in Science (iCons) program is based at its Integrated Science center, a dedicated building for fostering interdisciplinary research and teaching in science. The iCons program began in 2014 and is organized around two theme areas—Renewable Energy and Biomedicine and Biosystems—aligned with collaborative research strengths at UMass Amherst. Our source for detailed information on the iCons program was Scott Auerbach, the founding and current program director for the iCons program.

The iCons program begins in the Spring semester of the first year, with a course entitled "Global Challenges, Scientific Solutions," which trains students to excel in diverse teams on current, real-world problems of the students' choosing. In the second year, students take prerequisite courses in math and biology and other disciplines, preparing them for the iCons 2 course titled "Integrated Scientific Communication" which trains students communicate effectively about Energy and Biomedicine to both technical and non-technical audiences, with a discussion of societal needs as part of the presentation. A third year iCons3 course entitled "Team Discovery Lab" develops lab attitudes, knowledge, and skills relevant to real-world problems in multiple theme areas. iCons concludes in the fourth year with a year-long capstone, and 2 credits of "Integrative Team Science Seminar" that fosters reflection about the advanced study and the student's career in college. The iCons program is notable both for its vertically-integrated curriculum, and its emphasis on both preparing science practitioners and policy makers.

The iCons program reaches about 60 students each year in an institution with 22,000 undergraduate students. The iCons courses are designed with active learning and a focus on case studies and student-initiated research questions. The beginning of the iCons course includes a "boot camp" stage for intensive training in communication skills to enable students to give persuasive speeches, and to create effective Powerpoint presentations and proposals. The later part of the course features an emphasis on experiment design, where teams of iCons students competitively develop proposals to conduct experiments and then implement selected projects in what is called a "Divergent Design Problem." In only six weeks, students are asked to go from idea, to research, to design, to the final assignment which includes a public debate open to all iCons students and faculty to assess the projects.

The intensity of the iCons course sequence is based on an awareness of the need for students to take charge of their learning, and an acceptance of the role of challenge and even failure in promoting deep student learning. For example, in the iCons 3 Energy Lab, students design and conduct two research projects within a semester, and in both cases are required to reflect on what they learned, and to explain any failures in detail, with a description of what they would do differently given a second attempt. iCons faculty often discuss how "constructive failure" can promote learning, given the proper scaffolding and reflection. The structure of the iCons course is visually summarized below in Fig. 6.7.

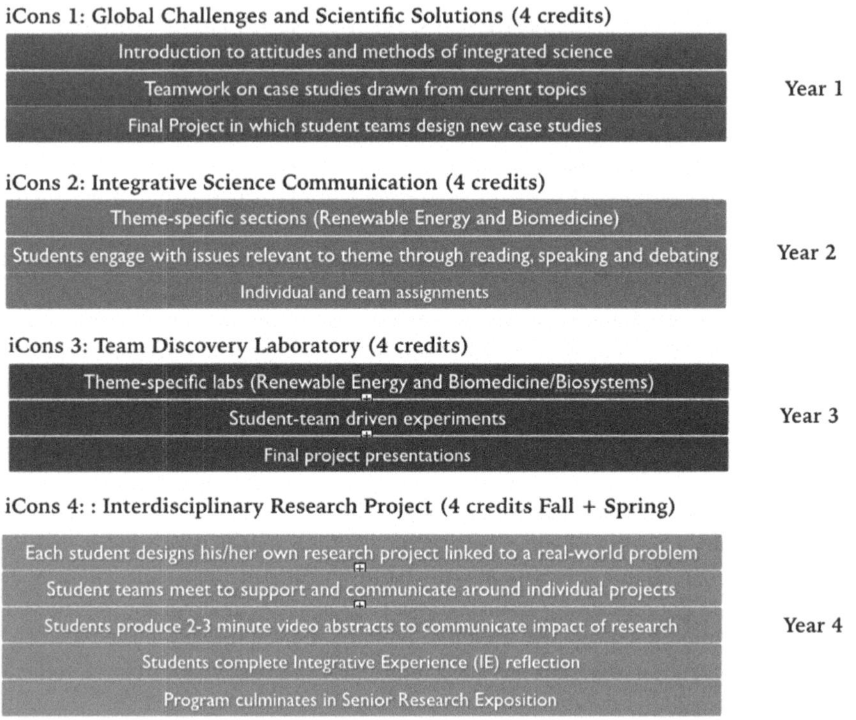

Fig. 6.7 Schematic of the UMass iCons Integrated Science curriculum (based on Auerbach 2015)

6.9 Princeton University Integrated Science

Princeton has traditionally offered one of the most demanding Integrated Science programs, that originally spanned three years of the undergraduate curriculum, and more recently has been condensed into a single year program. The course has a unifying theme of the mathematical models used to describe nature. Princeton Integrated Science is exceptionally rigorous and develops a very solid grounding in all aspects of quantitative modeling, including dynamical models, chemical kinetics, population growth, and also probabilistic models found in genomics and thermodynamics. In the first, year, Students get a full year of Chemistry, Physics and one semester of Computer Science and a substantial amount of Biology. The first-year course meets five days a week for an hour and has a 3 h lab and 3 h computational "precept" each week and counts as a double class.

Staffing of this ambitious course required a large number of faculty—in both the Fall 2010 and 2012 syllabus, 12 professors were listed, representing the departments of Physics, Ecology and Evolutional Biology, Computer Science, and Chemistry. Many or most of the faculty were based in the Lewis-Sigler Institute at Princeton, which is dedicated to "Integrative Genomics" and "Quantitative Biology," and the sequence leveraged the research activity and expertise of this institute. Students wrote their own programs and developed models using Java in the first semester, and by the second semester were working on advanced topics of computational biology. The large time investment from both faculty and students seemed to be worth it, and in the words of one of the Princeton students who completed the program, Michelle Ward '10:

> The course taught me to not limit my questions to a particular field. Rather, by asking questions about the chemistry and physics behind a biological system, I can learn about the biology from a more complete perspective. (Princeton ISC 2017)

The professors also felt strongly that including such a broad range of sciences provided stronger preparation for scientific research. In the words of David Botstein, one of the founders of the Integrated Science program, "Any budding researcher needs a foundation in several fields to be able to work on the most important problems confronting scientists today" (Botstein 2017; Bialek and Botstein 2004).

The current program has been restructured into a freshman year program that integrates computer science, biology, biochemistry and physics, with a very high level of mathematics applied throughout the curriculum. The intensive first year double course meets five times a week, as half of a student's four courses, and fulfils the introductory year-long course requirement in chemistry, physics, and semester credit for introductory majors-level courses in Computer science and molecular biology. Students can separately complete a certificate in quantitative biology, but this is no longer part of the Integrated Science curriculum. The program currently accepts 30–40 students who have high math proficiency (at the AP AB calculus level), and who are comfortable with the high level of mathematics.

6.10 Yale-NUS College (Singapore)—Integrated Science

Yale-NUS College (Singapore) was founded in 2013 with a mission to "redefine liberal arts and science education for a complex, interconnected world" (Yale-NUS College 2013). The Common Curriculum at Yale-NUS includes 10 different courses across the entire spectrum of disciplines and is designed to "integrate knowledge from across the disciplines and around the world." Science and Math courses within the Common Curriculum include a course in Scientific Inquiry, Quantitative Reasoning, and a sophomore science track. In the first three years of Yale-NUS College, the sophomore science courses included two tracks—Foundations of Science (intended for non-majors) and Integrated Science (intended for Science majors). The Integrated Science sequence included three semester-length courses, which included one semester in the first year of study, and a pair of courses in the second year.

The first of the Integrated Science courses was intended to merge all of the scientific disciplines within a framework of a common theme related to "The Science of Water" and was offered to the inaugural first two classes of students at Yale-NUS College in 2014 and 2015. The first semester Integrated Science course was intended to prepare students for science majors, by exposing them to a "big problem" based learning strategy, and to also include rigorous mathematics, computational science, biology, physics and chemistry. The course facilitators recognized that the breadth of the course and its expansive aims created some inevitable tensions which were hoped to be remedied by offering topics designed to pique scientific curiosity.

These topics were covered with the same level of depth as a disciplinary introductory course for majors, without the same extent of content coverage as a traditional disciplinary introductory course. To accomplish this goal, imaginative assignments were developed and a team-based teaching approach was used that included five professors within the classroom to teach approximately 37–40 students. All five of the "disciplinary content experts" were present in the classroom for each class—and represented the five disciplines mentioned earlier. Other pedagogical innovations included an online journal, where students wrote pieces on topics related to the course, a TEAL-style active learning classroom, and seamless integration between labs and lectures. A schematic of the course topics and applications is shown in Fig. 6.8.

The Integrated Science course was developed during a year before the Yale-NUS College curriculum was first offered in 2013–14. According to Shaffique Adam, one of the course leaders, the teaching team met (on average) for two hours each week for an entire year, and the earlier discussions were determining whether to use a "top-down" (math/physics) or "bottom-up (biology/earth-science) pedagogical approach. The theme that was finally chosen was "water" and viscosity, which allowed for ample ranges of examples and applications in the fields of Mathematics (Navier Stokes partial differential equation), Physics (fluid dynamics), Chemistry (chemical and

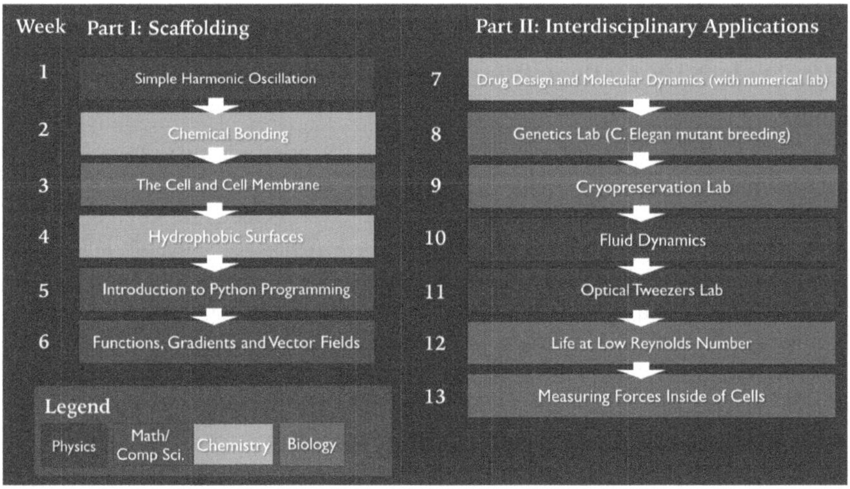

Fig. 6.8 Outline of the structure of the Yale-NUS College Integrated Science semester-long course in Science for science majors (based on figure from Shaffique Adams)

Hydrogen bonding) and Biology (life and circulation within organisms). Viscosity unified all of these fields, and included bacterial motion, hydrogen bonding, Reynolds number and other topics.

6.11 National University of Singapore—Special Programme in Science

One of the earliest curricular experiments at NUS with an Integrated Science curriculum was the Special Programme in Science (SPS), which began in 1996. SPS is a two-year interdisciplinary science curriculum designed to foster research skills in undergraduates in their first years at university. The program serves 80 students over two years, and involves four faculty who offer a mix of courses across Physics, Math, Life Sciences and Chemistry with a special emphasis on communication and programming skills. The program is structured around four key courses, that are sequenced in order of scale of the phenomena considered. The first course is "Atoms to Molecules" which covers physics and chemistry, followed by "The Cell" which is an introductory biology course, and then in the second year the courses called "The Earth" and "The Universe" cover earth science, and astronomy and astrophysics, respectively. Parallel to these thematic courses is the "Discovering Science" course in the first year, where students develop proposals for research projects, which they complete in the second year, and the "Integrated Science Project"—a research experience for second year students (Fig. 6.9).

As described by SPS Director Dr. Adrian Lee, SPS focuses on the skills practiced during research with a seamless integration of mathematics and communication skills in the curriculum. The SPS program makes extensive use of over 30 student mentors, who are able to supervise first and second year students in a group research project. The SPS program is also located in a dedicated space, which includes flexibly configured seminar/lecture rooms, a small library and computer lab, and a wet lab and a dry lab. The labs include instrumentation for cellular biology, as well as scanning tunneling microscopy, and spectrometers for physical science. The students in their

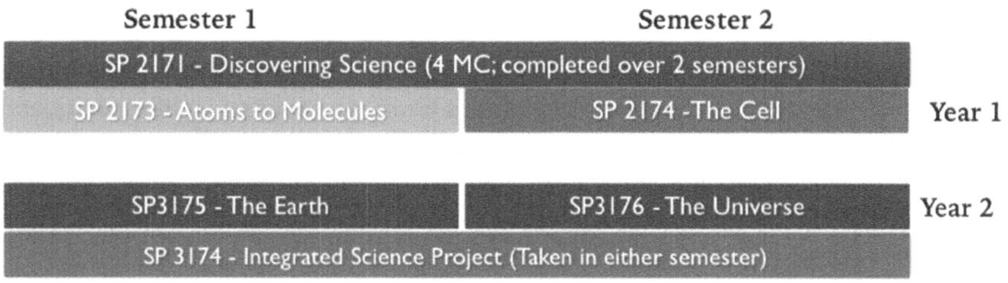

Fig. 6.9 Schematic of the NUS Special Programme in Science, showing the multi-year sequence of courses, culminating in an Integrated Science project, which is a research project chosen by students

classes are able to explore more advanced topics than in a typical introductory science curriculum, such as quantum mechanics, cosmology, relativity, and computer modelling. Students conduct laboratory experiments in the wet and dry labs, and also participate in a field lab experience in earth science, where they visit the NUS Marine Reserve and study the distribution of fauna on the beach at low tide. Since the SPS is designed to train research skills, in many of these lab exercises, students design their own research programs, such as a field research study at the Marine Reserve.

The key piece within SPS is the "discovery science module"—in which students propose a research project for the second year of the program. The course is completed over two semesters and goes by the title "Discovering Science." After students propose a project, a panel selects which projects are to be pursued, and each year about a dozen of these research projects are completed by students in groups of 2–3. These student research projects are supervised by a wide mix of faculty across NUS in multiple departments. These faculty mentors then become involved in the SPS assessment, and provide grades for a written report, as well as from a poster presentation and a group presentation of the results. The combination of written and oral presentation is part of the program design to enable students to develop more advanced communication skills.

Each course in the SPS sequence (Discovering Science, Atoms to Molecules, The Cell, The Earth, and The Universe) expands upon the previous course, with topics that are based on phenomena occurring within various distance scales in the universe. Preparing faculty for working in this environment requires active recruiting. In their recruitment, the SPS faculty look for others who have a strong sense of interdisciplinarity, which was assessed in some cases on the basis of their PhD research being based on interdisciplinary topics.

6.12 University of British Columbia, Science One

Science One is a single 27-credit course offered at the University of British Columbia (compared to a 36-credit full course load) offered to 75 first-year students by a team of 8 instructors. The Science One instructors include faculty from four disciplines, representing both include Educational Leadership track faculty and research-active faculty. Science One is described in its promotional materials as a "community, a place where you can immerse yourself in science." Science One is an intensive first-year introduction to science that integrates biology, chemistry, mathematics and physics. The Science One program includes a dedicated classroom and study area and makes use of the resources of UBC and adjacent field stations to provide "the highest level of first-year science offered at UBC." Examples of the high-level work includes a field trip to the UBC Bamfield Marine Science Center, a nearby marine biology research station, team research projects designed and conducted by students, and special guest lectures by a wide range of distinguished scientists. As described on the Science One web site, "Science One was founded on the idea of giving first-year students an advantage by giving them access to the best teachers and researchers."

The director of the program is able to recruit faculty from across the different disciplines for Science One, by stressing to the faculty the unique opportunity to learn how different disciplines approach subjects and teaching. Each discipline in Science One has 2 instructors—one "Educational Leadership" track instructor and one research-active faculty. The long track record of Science One makes it relatively easy for newer faculty to get to know the aims and approaches of the course. One key feature that helps the faculty share pedagogy and ideas is that the Science One teaching team is co-located in the same place with offices near a student study space. This co-location assures that the teaching team is easily able to see each other and to provide feedback on the teaching as they are all watching the classes together.

6.13 Yale-NUS College, Foundations of Science

The Yale-NUS College Foundations of Science (FOS) program was a year-long introduction to science within the original Yale-NUS Common Curriculum. Foundations of Science was intended to be the main science course for non-majors in science, and offered second-year students an interdisciplinary exploration of science that built upon the work students completed in the first year of the common curriculum. The author was one of the designers of this course and taught or led the course during all three years of its implementation and provided the information for comparing the course with the others in the study.

Within FOS each of the eight instructors in the teaching team offered a 5-week or 6-week "disciplinary case study" short course in their disciplines. Students took two of these disciplinary case studies in each semester from their instructors, and then groups of students were asked to work on interdisciplinary final projects. The entire teaching team met on a bi-weekly basis to plan and review the course. The scheduling of the course allowed for students to sample two of the four offerings during the semester, giving them exposure to four different disciplines through the year.

In the first year FOS was offered (2014–15), disciplinary case studies were merged with an immersive field trip, in which students visited Langkawi Island in Malaysia, and performed observations of ecology, geology, astronomy, and field biology in the field, using the rich natural ecosystem of Langkawi island as a laboratory. For the second year of FOS (2015–16), the disciplinary case studies were aligned with two themes—Evolution (gradual change as revealed by physics, chemistry and biology) and Revolution (sudden disruptive change due to disasters or disruptive technology). At the end of each semester, groups of four students from different sections were brought together to work on a Grand Challenge problem that allowed them to apply and integrate what they learned in their different disciplinary case studies. Half of the students were asked to "design a new disruptive technological innovation that can be used to deal with one of the consequences of global climate change" while the other half proposed a study to determine "what are some likely future adaptations of organisms and communities to the anthropocene".

The combination of a "deep dive" into two disciplines each semester and a chance to apply what was learned towards global climate change was a unique innovation for the course, which featured two 5-week case studies, and a 3-week interdisciplinary "Grand Challenge" exercise in each semester. The structure of the course is shown below in schematic, giving the tracks of two students who during the year would take four of the "deep dive" short courses and conduct two Grand Challenge exercises (Fig. 6.10).

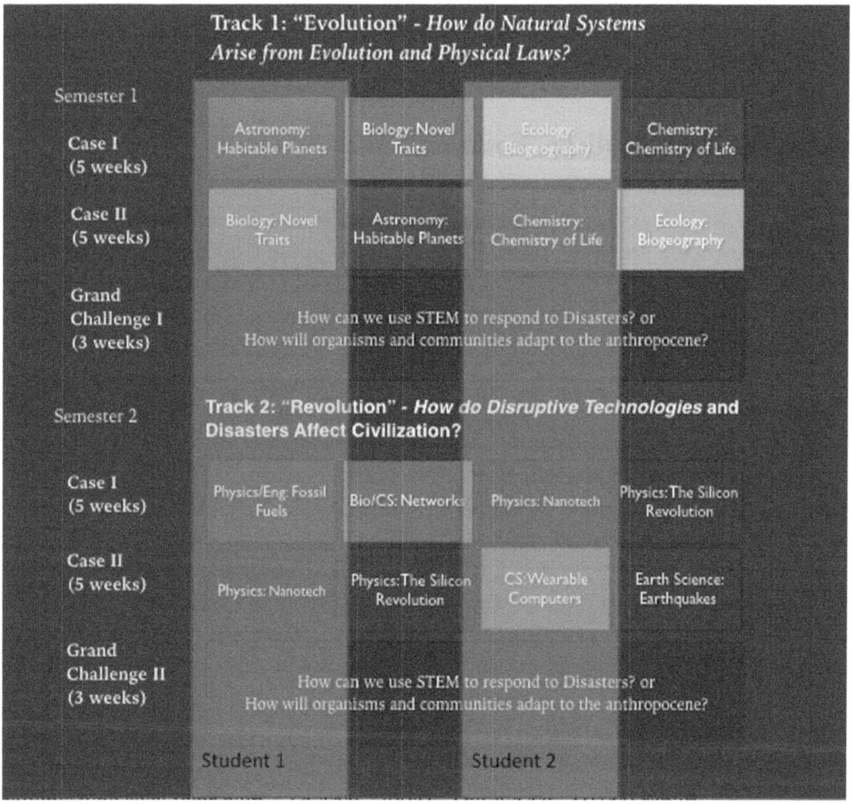

Fig. 6.10 Schematic of the Yale-NUS College Foundations of Science course, showing the way in which students would move through two "Disciplinary Case Studies" in each semester, followed by an interdisciplinary Grand Challenge problem students studied for three weeks

6.14 University of Hong Kong, Fundamentals of Modern Science

The Fundamentals of Modern Science course at Hong Kong University (HKU) is a course designed to provide a broad view of science that emphasizes the interconnections between disciplines, and surveys the history, fundamental concepts, and impact of science on civilization and society. The course arose from the recent (2012) restructuring of Hong Kong University curriculum that increased the duration of the undergraduate degree programs from three to four years. The extension opened opportunities for General Education courses in science such as Fundamentals of Modern Science. The course is organized around a theme of scale—and begins at the smallest spatial scales considering the fundamental structure of matter. After two weeks on particle physics, the course transitions toward atoms and nanoscience for 3.5 weeks, and then includes 3.5 weeks to explore biology at the scale of the cell. After the biology unit, topics in energy and heat are considered which leads toward a discussion of earth and its environment, and finally a two-week unit on Planet Earth and the universe to conclude the course.

The HKU Fundamentals of Science course is offered to 280 students each semester and is taught by a teaching team of 3 lecturers from diverse scientific disciplines, and a team of 3–4 tutors. The course meets 3 h per week for lecture, with an additional 2 h tutorial. Dr. Jason Pun, an astronomer by training, was interviewed to learn more about the motivations and development of the course. According to Dr. Pun, "We thought about a lot of different approaches but at the end we used the movie *Powers of Ten* as an inspiration to cover all different natural phenomena linked by the length scale." The quantum nature of matter was discussed, along with the Heisenberg Uncertainty principle, and then chemistry—with a strong emphasis on water and also carbon—as manifested in nanotubes, and other interesting nanotechnology contexts. Dr. Pun also pointed out that the transition to the largest scales of astronomy and cosmology links back at the end of the course to the smallest scales.

6.15 McMaster University, iSci Program

iSci at McMaster University is an "honours, interdisciplinary, research-based science program" designed to provide graduates with "high level research and communication skills" which as described in their own literature has been developed by an "unconventional group of collaborators, crossing discipline and administrative boundaries." Carolyn Eales, the iSci Program founder and former Director, and Chad Harvey, Associate Professor in the School of Interdisciplinary Science, were interviewed for to learn more about the history and vision for the program. The iSci program was envisaged in 2007, and began in 2009, and currently accepts 40–60 students each year, selected from over 450 applicants and is operated by 5 faculty and several staff members who create the instructional team. They employ a pedagogy known as "Research-based Integrated Education" which structures the learning around specific research problems. They provide these scaffolded research projects to "motivate learning" and to engage all of the scientific disciplines and develop independent research skills in students (Symons et al. 2017).

Among the topics studied in iSci are interplanetary exploration, pandemics, climate change, medical imaging, sustainable energy, environmental contamination, genetic modifications and "the complex issues facing society" that require "collaboration between multidisciplinary teams of scientists with the expertise to think creatively." The program progress through four levels in each of the four years. The program begins with a Foundations course, which study topics such as explanatory exploration, drugs, sustainable energy and cancer along with labs and science literacy training and additional electives. In the second year, iSci branches into a wider instructional team that develops six research projects with additional laboratory, science literacy and mathematics training. In the third year, the Level 3 program develops three more intense research projects followed by an independent project and concentration specific elective courses. Finally in Level 4 students complete their iSci Thesis projects and participate in a seminar. A strong emphasis on scientific literacy is developed throughout the four years and the program is integrated with the university library with a HQ in the Science and Engineering library and the head librarian co-leading the scientific literacy component.

The original faculty for iSci were selected specifically for their potential interest and ability for interdisciplinary work. Within the early group of faculty were also two faculty specifically recruited and assigned to teach in the program. Part of their preparation was an extended 8-month design exercise whereby the faculty "deconstructed" the first-year science program and together decided what each of the key essential concepts were from the various disciplines. iSci also incorporated staff members, such a librarian, into the teaching team. The deconstruction was followed by a construction using a deep look at potential research problems and an identification of the "key essentials" from each discipline (including knowledge, skills, experiences, and concepts). Two unifying themes, sustainable energy and planetary exploration, are integrated with the disciplinary essentials to build something of a "interdisciplinary quilt" that structures the first-year program.

6.16 Findings from Research into Interdisciplinary Science Teaching

The most effective of the interdisciplinary courses in our study were structured to arise naturally from themes or problems that were intrinsically interdisciplinary. Having a natural interdisciplinary integrating theme—such as global warming, unified processes over a range of spatial scales, or multifaceted problems within environmental science (water, biodiversity, etc.)—provided the necessary motivation for interdisciplinary work. In both the iCONs and AISS courses, "inherently interdisciplinary" contexts became drivers for curriculum development. In the best of the courses, these themes create a "community of practice" as it is the "artefact" within the community that becomes joint intellectual property (Martensson and Roxa 2014).

Preparing students and faculty for the challenges of the interdisciplinary environment is crucial. The majority of these programs are offered to a pre-selected "elite" group of students, which helps students rise to the challenge, but selectivity raises troubling questions about access and scale. Some general education interdisciplinary courses, such as the Yale-NUS Foundations of Science and University of Hong Kong Fundamentals of Science had no pre-selection, and attempted to serve the more diverse preparation. Such courses may need a pre-course "boot camp" experience to fully prepare students for the intense interdisciplinary environment and to shore up skills within the student population before the course begins.

The NUS SPS, iCONS AISS and UBC Science One programs are able to offer active learning and student-driven inquiry in their daily classes, and these principles have been successfully implemented by having the pedagogy drive the content chosen for the course. It became clear that once a strong theme for the interdisciplinary inquiry had been found, having pedagogy as a primary driver of content was a powerful tool in uniting faculty even as they come from different disciplinary backgrounds. The use of backward course design, with clearly articulated and assessable learning outcomes as a primary design principle, was also a powerful tool for unifying and clarifying the curriculum for both students and faculty. Using innovative assessment strategies, such as the iCONs self-assessment by students, enables the instructors to provide more agency within the student population for their learning, further increasing engagement within the students.

From analyzing the results from our interviews, we have identified a set of "best practices" in preparing faculty for teaching in multi-year interdisciplinary science programs, for collaborating and communicating grading and pedagogy within the course, and for assessing the outcomes of students. These practices were observed in multiple of the most stable and long-running interdisciplinary programs and were reported to improve the quality of the teaching and instruction and are summarized in Table 6.3.

To develop coherence and stability within the multi-year curriculum, most programs have developed extended planning and coordination mechanisms that include pre-semester retreats with the entire teaching team, detailed scheduling of sections and even individual class meetings in advance, and frequent discussions within the teaching team to monitor and refine the curriculum. Within each program are structures for administration that allow for multiple disciplines to be housed and work together, and for creating joint or dedicated appointments of faculty to the inter-

Table 6.3 Best practices within interdisciplinary science courses and programs

• Faculty selected based on interest and ability in interdisciplinary science; intensive orientation and planning involving all faculty, and junior faculty given additional mentoring and help developing interdisciplinary teaching
• Extended meeting of all program faculty before term supplemented by regular meetings to discuss curriculum and pedagogy. Cohesion from unifying themes and joint interest in developing excellent research with students. Co-location of offices and labs, and co-teaching in seminars and laboratories
• Common rubrics for grading combined with discussions among faculty, with some flexibility for disciplinary assessments built into programs. Peer evaluation and self reflection by students along with presentations graded by multiple faculty
• Discussing and monitoring performance of students in subsequent classes and in research; systematic examination of teaching evaluations and focus group discussions with students. Research productivity of students and alumni outcomes

disciplinary program (Fig. 6.11). In many cases, these programs employ dedicated postdoctoral fellows or junior faculty who can develop new approaches to teaching, with mentorship from more experienced faculty. Especially in multi-year programs, a significant investment of time is necessary for the course sequence to maintain its integration and coherence, to allow the faculty to understand the boundaries and overlaps in disciplinary expertise, and to find areas where synergies from blending disciplines can deepen student learning. This process is difficult but also enables these programs to engage deeply into problems where disciplinary boundaries can prevent effective problem-solving—thereby providing both a much-needed niche within science curriculum but a rewarding intellectual and academic environment that can attract and strengthen both our best science students and faculty as they can together develop new insights into vital issues that transcend disciplinary boundaries.

From the interviews with the leaders of the teaching teams, it is clear that interdisciplinary instructors must conduct frequent negotiation between disciplinary perspectives to give deeper insights—and this negotiation can be a very rewarding experience for both the faculty and for students. Regardless of how the courses are designed and maintained, the significant investment of time needs to be made to build a truly interdisciplinary curriculum, to allow the faculty to understand the boundaries and overlaps in disciplinary expertise, and to find areas where synergies from blending disciplines can deepen student learning. This process is difficult, but from our study it is clear that the deep perspectives and broad understanding of complex problems make interdisciplinary STEM education worth the effort.

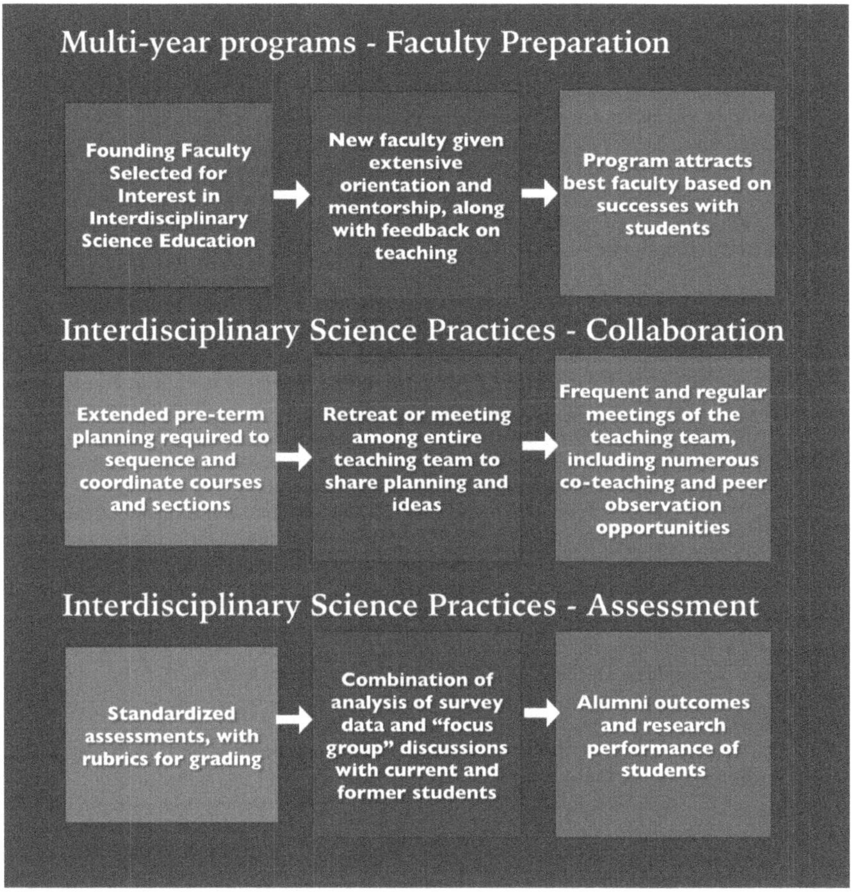

Fig. 6.11 Summary of "best practices" within multi-year interdisciplinary science programs, indicating some of the sequences of faculty preparation, collaboration and assessment found within programs in our study

References

AAC&U. (2010). Integrative and applied learning VALUE rubric, available at https://www.aacu.org/value/rubrics/integrative-learning.

AAAS. (2011, July). Vision and change in undergraduate biology education—A call to action. *Vision and Change*. Retrieved December 20, 2017, from http://visionandchange.org/.

AAMC. (2004, July). *Educating doctors to provide high quality medical care—A vision for medical education in the United States*. Report of the AAMC Ad Hoc Committee of Deans. Retrieved December 20, 2017, from https://members.aamc.org/eweb/upload/Educating%20Doctors%20to%20Provide%20July%202004.pdf.

AAMC. (2009). *Scientific foundations for future physicians*. AAMC-HHMI Committee Report, Washington, DC: Association of American Medical Colleges.

American Cancer Society. (2000, February 14). The role of the private sector in training the next generation of bioscientists. *Training the Next Generation of Biomedical Scientists*. Retrieved December 20, 2017, from https://www.bwfund.org/sites/default/files/media/files/Training%20the%20Next%20Generation.pdf.

Auerbach, S. (2015). *The iCons four-year curriculum plan*. Retrieved January 14, 2017, from http://icons.cns.umass.edu/sites/default/files/attachments/icons-at-a-glance-fall2015.pdf.

Beane, J. A. (1995). *Toward a coherent curriculum*. ASCD Yearbook, Association for Supervision and Curriuclum Development. Retrieved July 11, 2018, from https://books.google.com/books/about/Toward_a_Coherent_Curriculum.html.

Bialek, W., & Botstein, D. (2004). Introductory science and mathematics education for 21st-century biologists. *Science, 303*, 788.

Bloom, J. (2004). Patterns that connect: Rethinking our approach to learning, teaching, and curriculum. *Curriculum and Teaching* (James Nicholas Publishers), *19*, 5–26.

Botstein, D. (2017). *Integrated Science curriculum*. Retrieved March, 2017, from http://lsi.princeton.edu/integratedscience.

Bryk, A. S., Gomez, L. M., & Grunow, A. (2010). *Getting ideas into action: Building networked improvement communities in education*. Carnegie Foundation for the Advancement of Teaching, Stanford, CA. Retrieved July 11, 2018, from https://www.carnegiefoundation.org/resources/publications/getting-ideas-action-building-networked-improvement-communities-education/.

Chinese University of Hong Kong. (2017). *General Education Foundation (GEF) programme*. Retrieved December 21, 2017, from http://cas.nyu.edu/core/about-the-program/the-four-parts-of-the-core.html#scientific.

Copp, N. H., Black, K., & Gould, S. (2012). Accelerated Integrated Science sequence: An interdisciplinary introductory course for science majors. *Journal of Undergraduate Neuroscience Education, 11*, A76–A81.

Fogarty, R. (1991, October). Ten ways to integrate curriculum. *Educational Leadership*.

Global Research Council. (2016, May). Interdisciplinarity—Survey report for the global research council. *Annual Meeting*. Retrieved December 20, 2017, from https://www.globalresearchcouncil.org/meetings/annual-meetings/2016-meeting/.

Harden, R. M. (2000). The integration ladder. *Medical Education, 34*, 551.

Hubball, H., & Burt, H. (2015). An integrated approach to developing and implementing learning-centered curricula. *International Journal of Academic Development, 9*, 51.

Hubball, H., & Gold, N. (2007). The scholarship of curriculum practice and undergraduate program reform: Integrating theory into practice. *New Directions for Teaching and Learning, 112*, 5.

James Madison University. (2017, July 1). *B.S. in Integrated Science*. Retrieved December 20, 2017, from https://www.jmu.edu/bsisat/about/program-overview/index.shtml.

Lattuca, L. R. (2001). *Creating interdisciplinarity*. Nashville, TN: Vanderbilt University Press.

Malik, A. S., & Malik, R. H. (2011). Twelve tips for developing an integrated curriculum. *Medical Teacher, 33*, 99–104.

Martensson, K., & Roxa, T. (2014). Promoting learning and professional development through communities of practice. *Perspectives on Pedagogy and Practice, 5*, 1.

Mathison, S., & Freeman, M. (1998). The logic of interdisciplinary studies. *ERIC*. Retrieved December 20, 2017, from https://eric.ed.gov/?id=ED418434.

McMaster University. (2017). *Honours Integrated Science program*. Retrieved December 21, 2017, from https://www.science.mcmaster.ca/isci/.

National Research Council. (2000). *Addressing the nation's changing needs for biomedical and behavioral scientists*. Washington, DC: National Academy Press.

National Research Council. (2003). *Bio 2010—Transforming undergraduate education for future research biologists*. Washington, DC: National Academies Press.

Northwestern. (2017). *Integrated Science program*. Retrieved December 20, 2017, from http://www.isp.northwestern.edu/about/.

NYU Shanghai. (2017). *The five parts of the core*. Retrieved December 21, 2017, from http://cas.nyu.edu/core/about-the-program/the-four-parts-of-the-core.html#scientific.

Oliver, B, Jones, S., Ferns, S., & Tucker, B. (2007). *Mapping curricula: Ensuring work-ready graduates by mapping course learning outcomes and higher order thinking skills*. Presented at the Evaluations and Assessment Conference, Brisbane. Retrieved July 11, 2018, from http://c2010.curtin.edu.au/local/docs/paper3.pdf.

Pearson, M. L., & Hubball, H. T. (2012). Curricular integration in pharmacy education. *American Journal of Pharmaceutical Education, 76(*10), Article 204.

Princeton ISC. (2017). *Student profile*. Retrieved March 17, 2017, from http://lsi.princeton.edu/integratedscience/student-profiles/student-profile-michelle-ward-10.

Princeton University. (2017). *Integrated Science curriculum*. Retrieved December 20, 2017, from http://lsi.princeton.edu/integratedscience/.

Purvis-Roberts, K. L., Edwalds-Gilbert, G., Landsberg, A. S., & Copp, N. (2009). Accelerated Integrated Science Sequence (AISS)—An introductory biology, chemistry and physics course. *Journal of Chemical Education, 86*(11), 1295.

Richlin, L., & Essington, A. (2004). Overview of faculty learning communities. *New Directions for Teaching and Learning, 97*, 25.

Simkins, S., & Maier, M. (2009). *Just-in-time teaching: Across the disciplines, across the academy*. Stylus Publishing.

Symons, S. L., Colgoni, A., & Harvey, C. (2017). Student perceptions of staged transfer to independent research skills during a four-year honours science undergraduate program, *8*(1), 1.

The University of British Columbia. (2017). *Science One*. Retrieved December 21, 2017 from https://ScienceOne.ubc.ca/.

UMassAmherst. (2017a). *iCons program*. Retrieved December 20, 2017, from https://icons.cns.umass.edu/academics.

UMassAmherst. (2017b). *Why iCons?* Retrieved December 22, 2017, from http://icons.cns.umass.edu/academics/why-icons.

University of Hong Kong. (2017). *SCNC112 fundamentals of modern science*. Retrieved December 21, 2017, from http://cas.nyu.edu/core/about-the-program/the-four-parts-of-the-core.html#scientific.

University of Leichester. (2017). *The centre for interdisciplinary science*. Retrieved December 21, 2017, from https://www2.le.ac.uk/departments/interdisciplinary-science.

Virginia Tech. (2017). *Integrated Science curriculum*. Retrieved December 20, 2017, from https://www.ais.science.vt.edu/programs/isc.html.

Yale-NUS College. (2013). *Yale-NUS College common curriculum report*. Retrieved December 21, 2017, from http://www.yale-nus.edu.sg/wp-content/uploads/2013/09/Yale-NUS-College-Curriculum-Report.pdf.

Chapter 7
The Future of STEM Teaching and Learning

Abstract The twenty-first century has been described as the Fourth Industrial Revolution, and the emerging science and technology of our present century is described and compared with earlier industrial revolutions. In each case, transformative technologies have sparked revolutions that have shaped not only the economy, but also higher education and STEM education. The history of each of these revolutions is explored and the detailed ways that higher education and STEM education has evolved is reviewed in the past and present centuries. The main technologies of the current industrial revolution, Artificial Intelligence, Biotechnology, Nanotechnology and the Internet of Things are described in separate sections. The role of these emergent technologies in a new kind of STEM curriculum is described, along with examples of new courses based on these technologies at Stanford, MIT and Harvard. These new exponential technologies and the emerging realities they will bring challenge the sustainability of our planet and our notions of humanity, and how this may shape our STEM education and our universities in the coming century is described.

Keywords Fourth industrial revolution · Artificial intelligence · Synthetic biology · Future technology · Future curriculum

7.1 The Fourth Industrial Revolution and STEM Education

The Fourth Industrial Revolution (FIR) as a metaphor for the twenty-first century has been used extensively within business and high-level policy circles, but is less well known among science educators. Recent white papers describe how the FIR will "shape the future of education, gender and work" and have been the basis of a number of initiatives for planning for more advanced types of workers in the future economy (World Economic Forum 2017a, b). The FIR is based on the notion that the evolution of technology brings about a driving force that transforms manufacturing with impacts that ripple through society. The First Industrial Revolution arose from harnessing water and steam power and transformed manufacturing, urban landscapes and the nature of work. New sources of power and energy over the past two centuries have resulted in additional industrial revolutions which have also produced seismic changes in society and in education. The FIR is a useful frame for discussing the

B. E. Penprase, *STEM Education for the 21st Century*,
https://doi.org/10.1007/978-3-030-41633-1_7

future of STEM education in the context of rapid change in society and the world. The most distinctive features of the FIR are so-called "exponential technologies" which change rapidly and in an accelerated pace. These exponential technologies include biotechnology, artificial intelligence, the internet of things and new nanomaterials which necessarily will require a fundamental shift in our approaches to STEM education.

7.2 The First Two Industrial Revolutions and Education

To gain more insight into this model it is helpful to take a closer look at the previous industrial revolutions which began with steam power, which made huge impacts in the late 1700 s. As one author puts it, steam power was "the hub through which the spokes of coal, iron and cotton were linked" (Rosen 2010). The changes brought about from steam-powered manufacturing created the First Industrial Revolution which was described in an 1884 work by Arnold Toynbee entitled *Lectures on the Industrial Revolution* (Weightman 2007). Toynbee described how the expansion of power and mechanical production combined with a "political culture which was receptive to change" to create seismic shifts in financial arrangements, the workforce, and the larger society. As one author put it, "the Industrial Revolution is not merely an acceleration of economic growth, but an acceleration of growth because of, and through, economic and social transformation" (Hobsbawm 1968).

During and after the First Industrial Revolution, higher education experienced transformation as well. The leading educational institutions began to question the value of the traditional classical education, and developed what were then revolutionary notions such as general education and the introduction of the college elective. Harvard President Charles W. Eliot named this system "The New Education" (Eliot 1869), which eventually replaced the previously dominant format of Latin and Greek classics education eloquently outlined in the Yale report of 1828 (Yale Corporation 1828). Simultaneously, graduate education within the United States and across the world emulated the German university model for postgraduate research, which enabled the rise of dozens research universities within the United States.

The Second Industrial Revolution began with the next new power source—electricity—which similarly transformed manufacturing and urban life toward the end of the nineteenth century. The many inventions of that period launched what some have described as a "new economy" (Atkeson and Kehoe 2007) during the period from 1870 to 1900. With this new economy came a dramatic expansion of access to higher education and the proliferation of multiple types of higher education institutions in the United States and Europe (discussed in Chap. 1). Private philanthropy, fuelled by the immense profits from new industries such as railroads, oil, and steel, enabled the founding of multiple major universities, including Stanford University (1885), Vanderbilt University (1873), the University of Chicago (1890). Numerous small colleges were also founded in this period, such as Pomona College (1887), University of Southern California (1880), and a small technical institute known as the Throop College (1893), which later became Caltech. These new universities were responding to the dramatic societal and economic changes occurring at the end of the nineteenth century, and since most of them were co-educational, they also helped provide greater access for women for careers in STEM fields.

Further seismic shifts in US higher education came after World War II, which included massive investments in education by the federal government through the GI bill and the creation of community colleges in 1947, which began to increase the diversity of students attending university, as mentioned in Chap. 1. The influential *Science: The Endless Frontier* (Bush 1945) resulted in the creation of the National Science Foundation in 1950, which drastically increased the resources available for university scientists, and set the stage for the research-based disciplinary science education within US higher education for decades to come. Within six years, Federal funding for STEM subjects increased from $6 billion per year to over $35 billion between 1960 and 1966. By developing its extensive system of 3600 universities and colleges, the United States created a massive educational economy which enrolls more than 19 million students annually, grants nearly 3 million degrees and employs more than 3.6 million people, including 2.6 million faculty (Gregorian 2014).

7.3 Educational Responses to the Third Industrial Revolution

The most recent Industrial Revolution is generally attributed to the advance of computers in the late twentieth century—which is not a power source like electricity and steam—have which nevertheless has revolutionized society and work. The Third Industrial Revolution also has accelerated the expansion of higher education with the proliferation of online education (discussed in Chap. 5), greatly increased diversity on campuses, which includes far more international students and non-white students, and a massive globalization of academic research enabled by easier communication via online technologies. This intensified commitment to large scale higher education across the world has resulted in massive improvements in access to higher education within previously unreached populations in India, China as well as within the United States.

The Third Industrial Revolution enabled the development of massive online open courses which has expanded access to university education to millions of previously unserved students across the world. The MOOC has yet to replicate the best of in-person "high-touch" higher education (see Chap. 5), but new online technologies and courses will continue to improve and soon will enable students to rapidly build skills and knowledge asynchronously on and off campus, with adaptive courses matched to their skills and interests. These technologies can enable both research universities and liberal arts colleges to more efficiently teach students with diverse backgrounds, and to help expand their campuses to connect with a global community of both faculty and students. Small liberal arts colleges are working together to improve on-campus experience for students through online math courses for incoming students, language courses taught via videoconferencing, and new ways of merging social media with small class seminars. One example of an initiative of this sort is the Liberal Arts Consortium for Online Learning (or LACOL), which has brought together some of the leading US liberal arts colleges to explore these technologies (LACOL 2017).

The Third Industrial Revolution has brought educators to an environment where access to information is immediate and free, shifting the focus toward "active learning" pedagogies (see Chap. 2) that place a premium on collaboration within diverse teams in a project-based and peer learning environment (Mazur 2009). Many of the most thoughtful responses to "reform" in STEM education in recent years have responded to this shift in emphasis away from content mastery and analysis and toward greater emphasis on liberal arts and interpersonal skills imbedded within a more interdisciplinary curriculum. Examples include the Project Kaleidoscope Science initiative (Elrod and Kezar 2016), *Liberal Studies in Engineering* (Bucciarelli and Drew 2015), the American Physical Society SPIN-UP project (Hilborn et al. 2003), the HHMI *Scientific Foundations for Future Physicians* report (AAMC 2009)—which all emphasize more interdisciplinary approaches in STEM that develop student capacity for collaboration and social interaction within STEM courses and curriculum.

The Third Industrial Revolution, like its predecessor, also has inspired and enabled the founding entirely new institutions of higher education. Such twenty-first century universities and colleges typically emphasize a more global and interdisciplinary curriculum and a greater emphasis on collaborative work between students within a residential living context. Examples include Olin College of Engineering (discussed in Chap. 4), with its emphasis on project-based learning and its innovative engineering curriculum, Yale-NUS College in Singapore, with its expansive and interdisciplinary common curriculum (Rosenberg 2017), Duke-Kunshan University in China, with its notion of "rooted globalism" which blends an appreciation for a local culture with a wide-ranging exploration of international approaches to identity and society (Godwin and Pickus 2017), and Soka University of America, which develops student capacity to become "global citizens" and which is developing an entirely new Life Science program to prepare them to contribute to emerging fields in biology and biomedical research (SUA 2018).

7.4 Emerging Realities from the Fourth Industrial Revolution

The Fourth Industrial Revolution often is described as the result of an integration and compounding effects of multiple "exponential technologies" that include Artificial Intelligence, Biotechnologies, and Nanomaterials. One example of a FIR advance which combines multiple of these technologies might be a bioreactor that uses new synthetic organisms (built from DNA created on a computer and "bioprinted) which is operated using AI-enabled robotic assembly lines using new nano-materials to provide immense improvements in the efficiency of production. This kind of future factory of the FIR lacks most of the elements we associate with "industry" such as large centralized factories staffed with massive labour forces within large corporations. Exponential technologies like AI, biotechnology, and nanomaterials, are all easily developed in small or large scale at low cost, and should decentralize and disrupt industry as much as steam, electricity and personal computers disrupted the past few centuries. Some have described the convergence of these "exponential" technologies as providing a "Singularity"—which will provide untold benefits to humanity, as humans "transcend biology" according to some authors (Kurzweil 2005). These advances promise to help develop great progress on some of the most vexing problems of the century, such as entirely new types of medicine, new sources of fuels and perhaps even solutions to global climate change. New forms of STEM education will be vital to enable students to participate in such advances, as well as to help mitigate against some of the dangers. We consider each of these exponential technologies with their promises and perils in separate sections.

7.5 Artificial Intelligence

Artificial intelligence benefits from the exponential increase in computer power and decreasing cost in storage which has been ongoing for the past 5 decades, which is often described as "Moore's Law." The doubling of CPU power every 18–24 months has enabled new supercomputers to reach computation speeds of 300 PetaFlops (1 PetaFlop $= 10^{15}$ floating operations per second) in the latest supercomputer known as Milky Way 2 (Peters 2017), which represents an increase in speed of more than a factor of 300,000 in just two decades. The TaihuLight supercomputer in China features 10 million CPU cores and is rated at 93 Petaflops, and is being used to simulate the evolution of the physical universe, tracking the motions and forces of ten trillion digital particles over billions of simulated years (Burgess 2017). Such supercomputers are also capable of simulating molecular dynamics, bonding and biochemistry within cells with unprecedented precision, bringing great advances to biology and biochemistry. In addition to scientific gains from advanced computing, the ability to find very large prime numbers has national security implications, as a central part of cryptography.

A recent National Academies report identified the many ways that AI will boost scientific research across a range of disciplines. AI is expected to play a major role in drug discovery, diagnosis and treatment, as AI agents become fully integrated into both biomedical research and analysis of diagnostic images. The AI algorithms will be able to use machine learning and advanced models of to predict the shapes of biomolecules, their physical properties, and efficacy in treatment. AI can also use machine learning to scan through millions of reactions and vast databases of results to isolate trends and effective treatments (National Academies 2018a). These burgeoning new AI technologies will also enable more responsive and human-like online learning environments, which might enable much more effective and efficient STEM education.

Along with promise, AI offers peril in the form of what have entitled "Artificial Super Intelligence" (Bostrom 2014; Yampolskiy 2016). AI currently can outperform humans on specific and highly constrained tasks such as in games like Go and Chess, as well as in massive sorting and computational tasks. As AI grows in its capabilities, it will begin to acquire more cognitive complexity, just like human students. Learning Theory and Learning Taxonomies (like those discussed in Chap. 3) can be applied toward Artificial Intelligence as it becomes more sophisticated. New forms of Machine Learning will create AI agents that can translate their training from specific tasks and domains into entirely new contexts, thereby expanding the AI "Zone of Proximal Development" into ever larger regions. As AI approaches and exceeds human capacity for higher level learning outcomes (such as in evaluating and creating new ideas) it will reach a level of Artificial General Intelligence or AGI. In some scenarios, when AI surpasses human intelligence, and achieves Artificial Super Intelligence or ASI, strong ethical, software and hardware limits will be needed to prevent catastrophic outcomes.

The development and use of safe AI systems in a regime of ASI will require a concerted effort, since advanced AI systems can utilize Machine Learning to dynamically rewrite the algorithms guiding the AI program, potentially removing the AI from the control of its human creators. This phenomenon has been called the Singularity Paradox, and arises because "the danger of super-intelligent machines is that they are too stupid to have common sense" (Yampolskiy 2016). Such machines could create potentially disastrous outcomes from unexpected conflicts between preprogrammed goals for the AI, and moral and ethical complexities that arise from the AI single-mindedly pursuing its goals. One example would be for an AI system that is programmed to maximize paperclip production destroying the earth as it single-mindedly converts everyting into paperclips and defies human attempts to stop it (Rogers 2017). Such "Recursively Self-Improving Artificially Intelligent Systems" could defy any attempt by humans to stop it by reprogramming itself. It is therefore imperative that any ASI system include a fail-safe method to preserve human oversight as the AI agent assesses and adjusts its own internal codes and objectives. This process is sometimes called "Artificial Intelligence Confinement" and many authors have described ways in which super-intelligent systems could use unexpected methods to evade control, including co-opting humans through social media and through other persuasion to assist the ASI system in escaping confinement.

7.6 Biotechnology

The Fourth Industrial Revolution could produce a permanent shift from nonrenewable energy resources toward renewable energy enabled by bio-technology breakthroughs. Biotechnology, like steam and electricity of earlier industrial revolutions, could fuel the 'bio economy' of the coming century (Philp 2017). The growing population of Earth will need more than 50% more food by 2050, within a context of reduced agricultural capacity from soil degradation and global climate change. The emergence of "bio refineries" to use genetically modified microbes to provide a wide variety of food components and chemicals help avert shortages in food supplies in the coming century. New kinds of bio refineries could make use of flexible food stocks such as cellulose, biomass, and simple sugars, enabling mass production of a diverse range of fuels, pharmaceuticals and food products at low cost and high scale. This new type of manufacturing would also bring about a permanent reduction the use of fossil fuels and greenhouse gas emission in the coming decades. New synthetic organisms can also be used for environmental mitigation by removing various toxic compounds from the environment such as dangerous metals within landfills (Das 2010). Start-up companies are already designing these new organisms using standardized synthetic biology "wetware." Synthetic organisms have also been used for the development of biological circuits and computers, and even building materials to be "grown" using living materials known as "bio-bricks" (Cameron et al. 2014).

Biotechnology enabled by the FIR may finally enable a technological solution to the environmental threats arising from the buildup of CO_2 and other greenhouse gases from the massive factories from our first two Industrial Revolutions. Some authors have predicted that global warming could render the earth uninhabitable through an increase of more than 10°, which would result in widespread crop failures and large fractions of the world's populations subject to heat exhaustion and potential death. The increases in temperature reduce agricultural productivity—by as much as 15% for every degree of warming (Wallace-Wells 2017). New technologies could help absorb excess CO_2 using both bio-engineered organisms, and new materials within buildings that can absorb CO_2.

The dangers of synthetic biology are also apparent, however—including the greatly enhanced ease for bioterrorism using synthetic organisms as part of a Biowarfare capability. A recent National Academies report *Biodefense in the Age of Synthetic Biology* cautions about how new technologies "expands the range of actors" capable of creating dangerous microbes and reduces the time required for deployment. The report expresses the most concerns about "re-creating known pathogenic viruses, making existing bacteria more dangerous, and making harmful biochemicals via in situ synthesis" (National Academies 2018b). Thankfully barriers still exist to these frightful possibilities, such as difficulties in "Booting" earlier life forms, and complexities in DNA synthesis and assembly. Precautions are being taken to enhance our ability detection of such activity and to develop pre-emptive biological remedies for such pathogens.

7.7 Nanotechnology and the Internet of Things

In the new manufacturing regime enabled by FIR technologies (sometimes called the Internet of Things or IoT) nearly anything can be designed on a computer and then "printed" on 3D printers that create objects in countless materials, or even biological tissues. This capability will allow humans to turn "data into things and things into data." Materials used can include the familiar thermoplastics found in traditional 3D printers to large scale construction materials to clumps of atoms 10 nm across

(Gershenfeld 2012). This capability will enable printers to construct entire buildings, build microstructures with incredibly precise tolerances or create of biological structures for implants or even transplants of organs.

Several authors have cautioned about the downside of the IoT—such as the possibility of the IoT running amok, with tiny self-replicating nanomachines offering a potential for uncontrolled "mass destruction" by consuming the earth's resources through what was called "grey goo" (Drexler 1986). One more realistic concern about 3-D printing is that it enables printing of polymer guns, which is difficult to control with a democratized 3-D printing capability (Gershenfeld 2012). The IoT will also bring about a proliferation of smart devices, cameras, and internet connected sensors found in a modern "smart city" which may be vulnerable to hackers in ways that could threaten urban infrastructure. A new initiative known as the "Secure Internet of Things Project" project is working to develop more secure cryptographics and security mechanisms for IoT devices, new open-source hardware and software standards in operating systems and applications to reduce chances for device malware and malfunctioning, as well as more secure wireless networks.

7.8 Future FIR STEM Curriculum

Substantial changes to the science and technology curriculum will be needed to allow for students to enable the rapidly emerging areas of genomics, data science, artificial intelligence, robotics and nanomaterials to create a more sustainable environment and help mitigate the potential problems discussed earlier. Such a FIR STEM curriculum would reconsider the curriculum within the traditional "primary" sciences—biology, chemistry and physics, and place a higher premium for training in computer science subjects as a form of FIR literacy. Within biology, new approaches might include training within introductory courses to discuss emerging areas such as synthetic biology and molecular design. Some examples of reshaped Life Science curriculum can be found at Stanford University, where a new Problem Solving in Biology course has students design cures to real-world pathogens such as Lyme disease and HIV, using authentic data from scientific literature and experiment design (Cyert 2017), or a new course in Engineering Biology that allows students to design their own life forms on computers and "bio-print" them to solve practical problems such as curing diseases and environmental mitigation. These courses are a response to the emerging bio-economy which already exceeds $400 billion in the US alone (Endy 2016). Within the Stanford curriculum is now an entire new major known as Bioengineering, which trains students at the "interface of life sciences and engineering" and merges expertise and resources in the departments of Medicine, Biology and Engineering (Abate 2015).

Similar innovations within chemistry include a worldwide proliferation of courses and degree programs in "Green Chemistry" which blends chemistry, biology and environmental science to allow students to engage on real environmental problems such as synthetic fuels, bioplastics, toxicology, and to train students in techniques to reduce pollution (Mammino and Zunin 2015). New physics curriculum that emphasize FIR collaborative skills are being developed, where the curriculum is centred on projects where students design and build original musical instruments, cryptographic gadgets, and other inventions collaboratively (Perry 2013). Additional educational responses to the Fourth Industrial Revolution would be to retool STEM curriculum and institutions to provide new science programs and departments in emerging interdisciplinary fields to more efficiently provide trained workers to help advance and accelerate the development of ever-more sophisticated biotechnology, nano-technology materials, and artificial intelligence.

More than producing skilled workers, the new form of FIR STEM education will also produce informed technology citizens, equipped with new forms of literacy in genetics, computers, and materials, to enable them to make informed decisions about how these technologies can be ethically and sustainably harnessed. A new course "Law, Technology and Liberty" at the Stanford Law School, co-taught by law and bioengineering faculty members, is the sort of curriculum that will be needed to help foster new high-tech forms of informed citizenship. Drew Endy, one of the professors of this course, asks the questions—"what capacities should be available to all citizens? How many people should be able to read and write DNA? What should be the price of using a language?" as a way of framing the necessity for the democratization of advanced technologies, to insure civilization to succeed.

Blended instruction, and optimization of the "flipped" class, and online courses will make more efficient learning environments for students that can adapt for diversity in preparation within students. The Future of Education Report at MIT strongly emphasizes the need for leveraging online courses to strengthen the residential education for undergraduates, and to also give more flexibility and modularity of courses (MIT 2013). Examples of effective blended environments include the supremely popular CS 50 course at Harvard (Mendez 2014), the MIT introductory Electrical Engineering course, where course material is delivered entirely online with the in-person component focusing on laboratory and maker space time for students to build and test robots, and the MIT Circuits and Electronics course, which has been offered as an online course for residential students, who found the course to be less stressful, and who appreciated the ease of scheduling and additional speed for receiving feedback in their assignments (Roll 2017).

7.9 STEM and the Future Human Condition

The Fourth Industrial Revolution and its associated technologies such as biotechnology and artificial intelligence challenge some of our fundamental assumptions of what it means to be human, and our relationship with the natural world. How should STEM education respond to this new Human Condition? These humanistic concerns are inseparable from technical advancement, and a new FIR curriculum will need to reduce the divisions between humanities and STEM to create a more integrated "system" of education which can explore the newly emerging conceptions of self and identity within the FIR, including discussions of autonomy, free will, genetic vs social determinism. As smart AI-powered machines and other advanced technologies become more common within corporations, workers with traditional skills will be displaced, with a corresponding increase in the polarization of wealth and political influence. The convergence of the physical, digital and biological worlds will have profound effects on the planet, the economy and on society, and STEM education needs to adapt quickly to assure sustainability of our planet (Schwab 2016).

More than ever, higher education in the FIR age must develop capacity not just for analyzing and breaking a technical or scientific problem into its constituent parts, but must instead emphasize the interconnections between each scientific problem across global scales, and interrelations between physical, chemical, biological and economic dimensions of a problem. As one author has put it (Peters 2017), "there is a single planetary technical system" in which globally scaled markets enable "hundreds of thousands of transactions and information exchanges take place at the speed of light within the space of a microsecond." This speed can cause volatility and chaos in financial systems, and similar analogs of interconnected complex systems exist in the realms of marine ecology, forest conservation, global climate, GMO deployment in food production, and the effects of extinctions on the biosphere, to name a few examples. In all of these systems the rapidity of responses to the system and the larger network of interconnections can easily result in exponential responses to small perturbations, and the FIR curriculum needs to be train students to recognize and help manage the proliferating numbers of exponentially responding systems.

7.10 New Sequencing of Education to Renew Skills

The rapid pace of change within the FIR will require large-scale initiatives for updating skills after graduation and re-connecting alumni and more experienced workers with campus environments. Pathways for students to re-engage with their institutions after graduation will become imperative, not only to update the skills of alumni but to help younger students (and faculty) learn about the rapidly changing realities within the industrial and corporate sectors. One innovative initiative exploring new sequencing of higher education is the Stanford2025 project, which envisions several mechanisms whereby students can extend their education over longer timeframes. One model is the "open loop university" where students can experience six years of higher education over their entire adult careers that can allow them to blend their learning with life experience and provide value to the campus by returning as "expert practitioners" over several intervals—enabling students to refresh their skills while interacting with the campus community. Another model known as the "axis flip" that prioritizes skill development and competency training over content and disciplinary topics, requiring new methods of assessment resulting in a credential known as a "skill-print" that students would constantly renew and extend through their careers (Stanford2025 2013).

The hallmark of the FIR is exponential growth and rapid change, which gives the curriculum an imperative to update the content on an unprecedented frequency to match the rapid tempo of scientific and technological advances. A more responsive curriculum of this sort places an extremely high premium on faculty development and curriculum renewal, as well as the mandate to create students that can think and reinvent themselves within the changing world they will graduate into. Within future universities and colleges, both students and faculty will never be "done" with their educations, but instead must engage constantly with their colleagues and outside expertise and experience to frequently renew and update their skills. To enable faculty to be able to maintain current knowledge, more active and creative forms of faculty development will also be required, and the campus must become a constantly renewing collaborative hub of activity to maintain itself within the fast-paced FIR environment of the future.

7.11 Conclusions

The first three Industrial Revolutions provided evidence for the profound shifts in society, the economy and in education which resulted in a proliferation of curricular innovation and the establishment of new educational institutions. As in the previous Industrial Revolutions, the most profound effects of this Fourth Industrial Revolution on our society will not be realized for many decades. Unlike previous of these Industrial Revolutions, however, the FIR features the impacts of several compounding "exponential technologies" which all share the capacity for rapid increases in scale and reductions of cost. This rapidity of advance in technologies demands a more proactive response from the educational sector than the more gradual societal evolution and subsequent response from educational institutions present in earlier Industrial Revolutions.

The impacts of the emerging FIR technology in economic and environmental terms alone will require a drastic reconsideration of the curriculum within higher education, to enable students to comprehend both the individual technologies in detail but to be able to thoughtfully analyze and predict the evolution of networked systems of technology, environment and political systems. The dynamic responses with networked systems—and the exponential feedback effects can amplify the pace of change, as has been seen in the context of global climate change, and in many other physical and biological contexts. The STEM curriculum will need to focus on emerging technologies—robotics, AI, IoT, nanomaterials, genomics and biotech—to provide not only a workforce capable of developing new applications and products, but capable of interpreting the effects of these technologies on society, and using their training to provide sustainable and ethical uses of science and technology.

Our Colleges and Universities owe it to these students and our future to develop more interactive forms of pedagogy at all levels, and to embrace a curriculum that stresses perspectives from multiple disciplinary and cultural perspectives over static swathes of disciplinary "content" and that cultivates creative discernment and communication over mastery of skills that are subject to obsolescence. Many of the emerging liberal arts institutions in the US and Asia, and new types CTE curricula are providing useful examples of how to implement this a new model of FIR higher education. Higher education needs to recognize the necessity of adapting these new

FIR forms of education to assure the sustainability of our environment and economy, as well as to sustain the relevance of higher education as a responsive and vital component of society's response to the FIR. Taken together, these new forms of FIR education will prepare both students and faculty for leadership roles in a world of rapidly accelerating change, with a curriculum that develops both technical mastery and a deep awareness of ethical responsibility toward the human condition.

References

AAMC. (2009). *Scientific foundations for future physicians*. AAMC-HHMI Committee Report, Washington, DC: Association of American Medical Colleges.

Abate, T. (2015, October 22). *New bioengineering major culminated department's evolution*. Retrieved December 2, 2017, from https://engineering.stanford.edu/news/new-bioengineering-major-culminated-department-s-evolution.

Atkeson, A., & Kehoe, P. (2007). Modeling the transition to a new economy: Lessons from two technological revolutions. *American Economic Review,* 64–88.

Bostrom, N. (2014). Superintelligence: paths, dangers, strategies. Oxford, UK: Oxford University Press.

Bucciarelli, L., & Drew, D. (2015). Liberal studies in engineering—A design plan. *Engineering Studies,* 103–122.

Burgess, M. (2017, July 31). China's supercomputer simulated the universe with 10 trillion digital particles. *Wired Magazine*. Retrieved January 17, 2017, from https://www.wired.co.uk/article/china-simulate-universe-supercomputer.

Bush, V. (1945, July 1). Science—The endless frontier. *NSF*. Retrieved December 3, 2017, from https://www.nsf.gov/od/lpa/nsf50/vbush1945.htm.

Cameron, D., Bashor, C., & Collins, J. J. (2014). A brief history of synthetic biology. *Nature Reviews Microbiology,* 381–390.

Cyert, M. (2017). *Developing a new introductory biology curriculum*. Retrieved November 2, 2017, from https://vptl.stanford.edu/spotlight/developing-new-introductory-biology-curriculum.

Das, N. (2010). Recovery of precious metals through biosorption—A review. *Hydrometallurgy, 103,* 180.

Drexler, K. E. (1986). *Engines of creation*. Garden City, NY: Anchor.

Eliot, C. L. (1869). The new education. *The Atlantic, February 1869.*

Elrod, S., & Kezar, A. (2016). *Increasing student success in STEM: A guide to systemic institutional change*. Washington, DC: AAC&U.

Endy, A. (2016, April 27). *Yale-NUS college STEM innovation conference*. Retrieved December 2, 2017, from http://steminnovation.sg/wp-content/uploads/2017/06/Endy_Yale_NUS_STEM_v1.pdf.

Gershenfeld, N. (2012). How to make almost anything. *Foreign Affairs, 91*(6), 43–57.

Godwin, K., & Pickus, N. (2017, November 1). Liberal arts & sciences innovation in China: Six recommendations to shape the future. In *Publications—Center for international higher education* (Boston College Center for International Higher Education ed.). Retrieved December 3, 2017, from https://www.bc.edu/content/dam/files/research_sites/cihe/pubs/CIHE%20Perspective/CIHE%20Perspectives%208_ENGLISH_13NOV2017.pdf.

Gregorian, V. (2014, Winter). American higher education: An obligation to the future. *Carnegie Reporter* (Carnegie Foundation), 82–91.

Hilborn, R. C., Howes, R. H., & Krane, K. S. (2003). *Strategic programs for innovations in undergraduate physics*. Project Report. College Park, MD: The American Association of Physics Teachers.

Hobsbawm, E. (1968). *Industry and empire—The birth of the industrial revolution*. New York: The New Press.

Kurzweil, R. (2005). *The singularity is near*. New York: Penguin.
LACOL. (2017, July 1). *Liberal arts consortium for online learning*. Retrieved December 3, 2017, from http://lacol.net/.
Mammino, L., & Zunin, V. G. (2015). *Worldwide trends in green chemistry education*. Cambridge: Royal Society of Chemistry.
Mazur, E. (2009). Farewell, lecture? *Science, 323*, 50–51.
Mendez, C. (2014, September 18). *This is CS50* (Harvard Crimson ed.). Retrieved December 2, 2017, from http://www.thecrimson.com/article/2014/9/18/this-is-cs50/.
MIT. (2013, July 1). *Institute-wide task force on the future of MIT education*. Retrieved December 2, 2017, from https://future.mit.edu/.
National Academies. (2018a). *Artificial intelligence and machine learning to accelerate translational research: Proceedings of a workshop in brief*. Washington, DC: The National Academies Press. Retrieved from https://doi.org/10.17226/25197.
National Academies. (2018b). *Biodefense in the age of synthetic biology*. Washington, DC: National Academies Press. Retrieved January 17, 2019, from http://nap.edu/24890.
Perry, C. (2013, September 23). *In Ap 50, students own their education*. Retrieved December 2, 2017, from https://www.seas.harvard.edu/news/2013/09/in-ap-50-students-own-their-education.
Peters, M. (2017). Technological unemployment: educating for the fourth industrial revolution. *Journal of Self-Governance and Management Economics, 5*(1), 25–33.
Philp, J. (2017). The bioeconomy, the challenge of the century for policy makers. *New Biotechnology, 40*, 11–19.
Rogers, A. (2017, October 21). The way the world ends: not with a bang but a paperclip. *Wired Magazine*. Accessed January 9, 2020, available at https://www.wired.com/story/the-way-the-world-ends-not-with-a-bang-but-a-paperclip/.
Roll, N. (2017, June 15). *For-credit MOOC: Best of both worlds* (Insight Higher ed.). Retrieved December 3, 2017, from https://www.insidehighered.com/news/2017/06/15/credit-mooc-proves-popular-among-mit-students.
Rosen, W. (2010). *The most powerful idea in the world—A story of steam, industry and invention*. Chicago: University of Chicago Press.
Rosenberg, J. S. (2017, July–August). An educated core. *Harvard Magazine*. Retrieved January 14, 2019, from http://harvardmagazine.com/2017/07/an-educated-core.
Schwab, K. (2016). *The fourth industrial revolution: What it means and how to respond*. Geneva: Global Agenda, World Economic Forum.
Soka University of America. (2018). *General education curriculum*. Retrieved January 14, 2019, from http://www.soka.edu/academics/general-education-curriculum/default.aspx.
Stanford2025. (2013, June 1). *Learning and living at stanford—An exploration of undergraduate experiences in the future*. Retrieved December 3, 2017, from http://www.stanford2025.com/.
Wallace-Wells, D. (2017, July). The uninhabitable earth. *New Yorker*.
Weightman, G. (2007). *The industrial revolutionaries*. Grove Press.
World Economic Forum. (2017a). *Accelerating workforce reskilling for the fourth industrial revolution an agenda for leaders to shape the future of education, gender and work*. Geneva: White Paper, World Economic Forum.
World Economic Forum. (2017b). *Realizing human potential in the fourth industrial revolution—An agenda for leaders to shape the future of education, gender and work*. Geneva: White Paper, World Economic Forum.
Yale Corporation. (1828). *Reports on the course of instruction in Yale College*. New Haven, CT: Yale Corporation.
Yampolskiy, R. V. (2016). *Artificial superintelligence: a futuristic approach*. Taylor and Francis: Boca Raton.

The manufacturer's authorised representative in the EU is Springer Nature Customer Service Centre GmbH, Europaplatz 3, 69115 Heidelberg, Germany. If you have any concerns regarding our products, please contact ProductSafety@springernature.com

Printed and bound by CPI Group (UK) Ltd, Croydon, CR0 4YY
30/09/2026
02225608-0011